Structural Pattern Recognition using Graph Matching

This book presents a comprehensive exploration of structural pattern recognition with a clear understanding of graph representation and manipulation. It explains graph matching techniques, unearthing the core principles of graph similarity measures, subgraph isomorphism, and advanced algorithms tailored to various pattern recognition tasks. It bridges the gap between theory and application by providing case studies, hands-on examples, and applications. It is a reference book for academicians, researchers, and students working in the fields of structural pattern recognition, computer vision, artificial intelligence, and data science.

- Begins with the fundamentals of graph theory, graph matching algorithms, and structural pattern recognition concepts and explains the principles, methodologies, and practical implementations
- Presents relevant case studies and hands-on examples across chapters to guide making informed decisions by graph matching
- Discusses various graph-matching algorithms, including exact and approximate methods, geometric methods, spectral techniques, graph kernels, and graph neural networks, including practical examples to illustrate the strengths and limitations of each approach
- Showcases the versatility of graph matching in real-world applications, such as image analysis, biological molecule identification, object recognition, social network clustering, and recommendation systems
- Describes deep learning models for graph matching, including graph convolutional networks (GCNs) and graph neural networks (GNNs)

Structural Pattern Recognition using Graph Matching

Approximate and Error-Tolerant Algorithms

Shri Prakash Dwivedi
Ravi Shankar Singh

CRC Press
Taylor & Francis Group
Boca Raton London New York

CRC Press is an imprint of the
Taylor & Francis Group, an **informa** business
A CHAPMAN & HALL BOOK

First edition published 2026
by CRC Press
2385 NW Executive Center Drive, Suite 320, Boca Raton FL 33431

and by CRC Press
4 Park Square, Milton Park, Abingdon, Oxon, OX14 4RN

CRC Press is an imprint of Taylor & Francis Group, LLC

ISBN: 978-1-032-85034-4 (hbk)
ISBN: 978-1-032-87360-2 (pbk)
ISBN: 978-1-003-53224-8 (ebk)

DOI: 10.1201/9781003532248

Typeset in CMR10 font
by KnowledgeWorks Global Ltd.

Publisher's note: This book has been prepared from camera-ready copy provided by the authors.

To my parents
and my family.

Contents

List of Figures

List of Tables

Preface

In recent years, pattern recognition has undergone a revolutionary shift, embracing the power of graph theory to enable more powerful and versatile data analysis. Graphs offer a natural and intuitive way to represent complex relationships among data points, making them invaluable for recognizing patterns in various applications, including image recognition, bio-informatics, social network analysis, and many more. Bridging the gap between theory and practical applications, this book aims to explore the exciting world of structural pattern recognition and graph matching techniques.

Comprehensive in its scope, this book provides readers with a thorough introduction to the foundational concepts of graph theory, graph matching algorithms, and structural pattern recognition. It delves into the various ways in which data can be represented using graphs and explores the challenges involved in matching and comparing graphs.

At the heart of this book lie innovative graph matching techniques that harness the representational power of graphs to address structural pattern recognition challenges across diverse fields. Its contents encompass a wide array of exact, approximate, and error-tolerant graph matching algorithms. Given the widespread utilization of graphs in engineering and science for representing relationships between objects, graph matching has become pivotal in object identification through various graph-based representations. However, the lack of efficient graph matching solutions has prompted the development of numerous approximation and suboptimal algorithms to address the complexities involved.

With meticulousness and clarity, the book presents an extensive survey of exact and error-tolerant graph matching techniques. One notable approach introduces a category of graph matching algorithms that optimize efficiency by removing less relevant nodes, effectively reducing the size of the graph. Additionally, the book explores an alternative strategy that capitalizes on node centrality information to identify and eliminate a fraction of nodes in both graphs, thereby enhancing performance in alignment with specific application requirements.

Recognizing the inherent link between the geometry and topology of graphs, the book presents a novel method to measure graph similarity through geometric graphs. This inventive approach defines vertex and edge distances based on the position, angular orientation, and length of the graph's vertices and edges. The culmination of these distance measures results in a comprehensive definition of graph distance between two geometric graphs. By employing this graph similarity framework, the book facilitates error-tolerant graph matching without significantly altering the original graph topology.

The graph kernel provides us with a way to embed the space of graphs into an inner product space, utilizing the graph kernel similarity measure. By adopting ideas originally designed for statistical pattern recognition represented by feature vectors, we can enhance graph matching techniques. This book describes fundamental concepts and techniques for graph kernels and embedding. Graph matching is a powerful technique employed in image analysis to compare and align structural representations of objects or scenes, represented as graphs. The book delves into the methodologies, and applications of graph matching in image analysis. Social Network Analysis is a powerful tool for understanding the intricate relationships and interactions among individuals, groups, or entities within a social system. This book also explores the basic concepts, methodologies, and applications of graph matching in Social Network Analysis.

Beyond its technical merits, this book serves as an essential guide for those interested in the field of structural pattern recognition and graph matching techniques. It empowers readers to leverage the rich and expressive representation capabilities of graphs to uncover hidden patterns within their data. By inspiring exploration of innovative applications, this book equips researchers and practitioners alike with the knowledge and tools necessary to excel in this dynamic and rapidly evolving domain.

As an extended and revised version of the first author's PhD thesis, the book caters to a diverse audience. Its primary target comprises graduate students and researchers working in the fields of graph matching and pattern recognition. However, it extends its reach to encompass researchers, graduate students, and industry professionals in computer science, artificial intelligence, and data science. While assuming a basic understanding of graph theory and pattern recognition, the book strives to remain accessible to readers with varying levels of expertise, making it a valuable resource for anyone seeking to delve into the world of graph matching and its applications.

In conclusion, this comprehensive exploration of graph matching techniques opens up exciting opportunities for pattern recognition by exploiting the representational power of graphs. The book's blend of theoretical concepts and practical applications promises to be an enlightening and inspiring journey for readers, empowering them to uncover hidden patterns and drive innovation across various domains. With its diverse target audience in mind, this book stands as a testament to the growing significance of graph-based methodologies in modern data analysis and pattern recognition.

Author Biography

Dr. Shri Prakash Dwivedi is an esteemed researcher and academic in the field of Computer Science and Engineering, specializing in pattern recognition, graph matching, and algorithms. He is currently serving as an Assistant Professor in the Department of Information Technology at G.B. Pant University of Agriculture & Technology, Pantnagar, where he has been actively involved in teaching and research. Dr. Dwivedi obtained his Ph.D. in Computer Science and Engineering from the Indian Institute of Technology (BHU), Varanasi, where he worked on graph matching algorithms for structural pattern recognition. Prior to that, he earned his M.E. in Computer Science from the Indian Institute of Science (IISc), Bangalore. With a prolific research portfolio, Dr. Dwivedi has published extensively in leading journals and conferences, including Pattern Recognition Letters, Springer's Lecture Notes in Computer Science, Lecture Notes in Electrical Engineering, ACM, and IEEE. His work primarily focuses on error-tolerant graph matching, similarity measures, approximate algorithms, and computational methods in structural pattern recognition. Beyond academia, Dr. Dwivedi has industry experience as a Software Engineer (R&D) at various software companies, where he worked on scientific computing, algorithm design, and software development. He has also contributed to research projects integrating IoT and cloud-based automation systems. A recipient of several accolades, Dr. Dwivedi secured All India Rank 1 in GATE 2006 and has been a holder of the MHRD Scholarship for his academic excellence.

Dr. Ravi Shankar Singh is working as a Professor in the Department of Computer Science and Engineering, Indian Institute of Technology (BHU) Varanasi, India and has an academic background of B.Tech. (CSE), M.Tech. (CSE) and Ph.D.(CSE). He has been a Senior Member of IEEE and ACM. More than 50 research papers have been included in his account. His research interests include high-performance and cloud computing.

Symbol Description

Symbol	Description
G	Graph.
GED	Graph Edit Distance.
$d_{a,b}(i,j)$	Damerau-Levenshtein distance.
$c(e_i)$	Cost of edit operation e_i.
$\varphi(G,H)$	Edit distance paths from G to H.
$kNC(G)$	k degree node contraction on a graph G.
$rNC(G)$	r centrality node contraction on a graph G.
S_n	It is set of all assignments φ from V_1 of G_1 to V_2 of G_2.
E_{ij}^A	Angular orientation component of edge E_{ij}.
E_{ij}^L	Length component of edge E_{ij}.
E_{ij}^P	Position component of edge E_{ij}.
φ_n^P	Graph embedding.
VD	Vertex distance.
ED	Edge distance.
GD	Graph distance.

Chapter 1

Introduction

The graph is a fundamental mathematical structure within the domain of computer science, which holds a prominent status due to its wide range of applications. It is defined as a collection of nodes and edges, where each edge acts as a link connecting a pair of nodes. Its inherent ability to represent structural relationships has expanded its use across a multitude of disciplines in science and engineering.

In the field of biology, graphs serve as an invaluable tool for modeling biological networks. Nodes within these graphs stand as representations of essential biological entities, encompassing cells, proteins, neurons, and an array of other units, whereas edges represent the inherent connections and relations between these entities, offering a powerful framework to analyze the complex interplay within biological systems. From cellular signaling pathways to neural networks, this representation unlocks a deeper understanding of biological phenomena. Within the domain of chemistry, graphs have extensive use as molecular graphs. Here, nodes symbolize atoms or molecules, while edges denote chemical bonds or valences between these entities. This representation equips chemists with a powerful visual tool to understand the structural composition of molecules, enabling the development of novel compounds, catalysts, and materials with diverse applications across industries and healthcare.

Graphs permeate virtually every facet of computer science. For example, within operating systems, it plays a crucial role in characterizing resource allocation. Nodes denote processes and resources, while edges represent allocations. An edge extending from a resource to a process indicates allocation, whereas an edge from a process to a resource signifies a request for the corresponding resource. This model is fundamental for the efficient management of resources in multitasking environments. In the domain of computer networking, graphs are instrumental in selecting the shortest path for routing data packets across communication networks. This application is essential for ensuring optimal data transmission, minimizing latency, and maximizing network efficiency in the age of global connectivity and data-driven technologies.

Within the area of software engineering, control flow graphs provide valuable insights into program complexity. These graphs allow developers to visually map the flow of control within a program, enabling the identification of potential areas for optimization and refinement. They also serve as a means to understand the dependencies and associations between different components

DOI: 10.1201/9781003532248-1

of a software project, assisting in the design and maintenance of robust and scalable software systems.

Graphs provide a natural and intuitive framework for representing complex relationships among data points, making them indispensable tools for pattern identification across a wide range of domains, such as image analysis, bioinformatics, social network analysis, and beyond. The objective of this book is to study the representational power of graph and apply it to structural pattern recognition and graph matching. Bridging the gap between theory and practical applications, this book aims to explore the exciting world of structural pattern recognition and graph matching techniques.

This chapter is structured as follows. Section 1.1 provides an introduction to graph matching. Section 1.2 describes the basics concepts of graph theory including the various representational techniques, and Section 1.3 presents an introduction to pattern recognition. Finally, Section 1.4 describes the organization of this book.

1.1 Introduction to Graph Matching

Graph matching comprises the process of evaluating the similarity between two graphs. This concept builds upon tree matching, which itself extends from string matching. Unlike strings, which are linear structures and allow for the appending of only one element at a time, trees and graphs can connect to multiple data elements simultaneously. This increase in representational power, from strings to trees and then to graphs, brings about a corresponding rise in computational complexity for matching algorithms, as depicted in Figure 1.1. While polynomial-time algorithms exist for string and tree matching, no such algorithm has been discovered for graph matching.

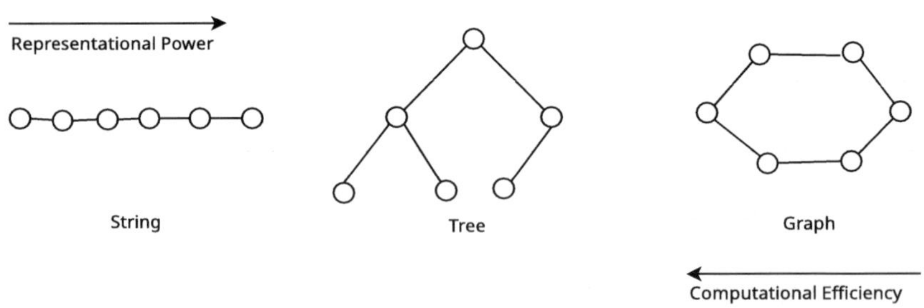

FIGURE 1.1: Generalization from string to tree and tree to graph

Based on the nature of the matching process, graph matching is usually classified into exact and inexact matching. Exact graph matching involves establishing a precise one-to-one correspondence between the vertices and edges of two graphs. It is similar to the graph isomorphism problem, where a bijection is required between the vertices of one graph and another such that for every edge in the first graph, there exists a corresponding edge in the second graph connecting the same set of nodes. While exact graph matching has theoretical importance, it may not always be practical for real-world applications. It is because input data often undergo modifications due to noise and errors in storage and transmission processes. Inexact graph matching (also known as error-tolerant graph matching) addresses this challenge by specifying a similarity score between two graphs. It accommodates a degree of tolerance to noise or errors that may have occurred during data processing. This approach proves vital in scenarios where some degree of flexibility in matching is necessary.

One of the primary applications of graph matching lies in structural pattern recognition. This field leverages the inherent structure of objects to perform a wide range of pattern recognition tasks. The ability to identify patterns is a fundamental characteristic of not only humans but also various living beings to a certain extent. Pattern recognition necessitates the analysis and classification of patterns based on their distinctive characteristics and features. In our day-to-day activities, we encounter numerous instances of pattern recognition. These range from recognizing a specific individual in a group, locating a particular book on a shelf, to identifying streets and a friend's home. While the complex cognitive system of the human brain makes these tasks appear intuitive and straightforward, performing the same tasks using a machine or computer can be considerably complicated. Pattern recognition systems develop algorithms to automate these tasks using computational methods. This book delves into the intricacies of graph matching and its pivotal role in structural pattern recognition.

Despite the rapid progress in pattern recognition, the challenge of efficiently solving the graph matching problem persists. Efficient polynomial time solutions remain elusive, prompting the recent emergence of a variety of approximation and suboptimal algorithms. These innovations represent a significant step forward in the pursuit of more effective methods for matching patterns represented by graphs, thus expanding the applicability of structural pattern recognition in practical, real-world scenarios.

1.2 Basics Concepts of Graph Theory

Graph theory is a branch of mathematics that deals with the study of graphs, which are mathematical structures used to model relationships between objects. Graphs are widely applicable, finding applications in computer science, physics, chemistry, social sciences, biology, transportation networks, and many other fields. In this section, we explore the fundamental concepts of graph theory, providing a foundation for further study.

A *graph* G is defined as an ordered pair $G = (V, E)$, where V is a finite set of elements called vertices or nodes, and E is a set of ordered or unordered pairs of distinct vertices, called edges.

In an *undirected graph*, edges have no direction, meaning they simply connect two vertices. In a *directed graph* (or digraph), edges have a direction, indicating a one-way relationship between vertices.

A *weighted graph* associates a numerical value, called a weight, with each edge. These weights can represent various attributes like distances, costs, or capacities.

A *simple graph* is an undirected graph with no loops (edges that connect a vertex to itself) and no multiple edges (two or more edges connecting the same pair of vertices).

A *multigraph* allows multiple edges between the same pair of vertices. A pseudograph allows loops, meaning an edge can connect a vertex to itself.

A *directed acyclic graph* (DAG) is a directed graph with no directed cycles. These are particularly important in applications like scheduling, where tasks must be completed in a certain order.

An *adjacency matrix* is a two-dimensional array of size $|V| \times |V|$, where $|V|$ is the number of vertices. The entry $A[i][j]$ is either 1 or 0, depending on whether there is an edge from vertex i to vertex j. An *adjacency list* represents a graph as an array of lists, where each list corresponds to a vertex in the graph. The list contains all the vertices that are adjacent to the current vertex.

The *degree* of a vertex in an undirected graph is the number of edges incident to it. In a directed graph, we have in-degree (number of edges coming in) and out-degree (number of edges going out).

A *path* in a graph is a sequence of vertices where each adjacent pair is connected by an edge. A cycle is a path that starts and ends at the same vertex.

A *complete graph* is a simple graph in which every pair of distinct vertices is connected by a unique edge.

A *bipartite graph* is a graph whose vertices can be divided into two disjoint sets in such a way that no two vertices within the same set are adjacent.

1.2.1 Representing Data as Graphs

In the domain of computer science and information technology, representing data as graphs has emerged as a powerful paradigm for visualizing and analyzing complex relationships. This approach exploits the mathematical framework of graph theory to model connections between various entities, providing insights that are otherwise difficult to determine through traditional tabular or linear data structures.

At its core, a graph is a mathematical structure composed of nodes (or vertices) and edges. Nodes represent entities, while edges signify the connections or relationships between them. This abstraction allows for the depiction of intricate networks, making it a versatile tool in various domains.

In a graph, nodes encapsulate pieces of information or entities, which can range from individuals in a social network to cities in a transportation network. Edges, on the other hand, define the associations or interactions between these nodes. They can be directional, indicating a flow from one node to another, or undirected, signifying a mutual relationship.

An attributed graph is a type of graph in which additional information—called attributes—is associated with its nodes (vertices), edges (links), or both. These attributes can represent diverse forms of information such as numerical values, labels, symbols, or feature vectors that describe the properties of the entities (nodes) and their relationships (edges).

An attributed graph is typically defined as:

$$G = (V, E, \mu_V, \mu_E)$$

where:

- V is the set of vertices (nodes),
- E is the set of edges,
- μ_V maps each vertex to a set of node attributes,
- μ_E maps each edge to a set of edge attributes.

Key features of attributed graphs are node attributes and edge attributes. Node attributes describe individual node properties such as, label, category, weight, color, feature vector. Edge attributes encode information about the relationship between nodes, e.g., distance, similarity score, connection type, direction, weight. Attributes itself can be numerical, (e.g., age, intensity, cost) categorical (e.g., gender, class label, type) and vectorial (e.g., embeddings, feature vectors).

A labeled graph is a type of graph in which additional information—called labels—is associated with the elements of the graph, typically nodes (vertices), edges, or both. These labels can be categorical (e.g., names, types) or numerical (e.g., weights, costs), depending on the application.

A labeled graph is typically defined as:

$$G = (V, E, \ell)$$

where:

- V is a set of vertices,
- $E \subseteq V \times V$ is a set of edges,

- ℓ is a labeling function, which can be, $\ell V : V \to \sum V$ assigning labels to vertices, or $\ell E : E \to \sum E$ assigning labels to edges, or both, depending on whether the graph is vertex-labeled, edge-labeled, or fully labeled.

Here, $\sum_V$ and $\sum_E$ are sets of possible labels (also known as alphabets or label domains).

Labeled graphs can be vertex-labeled, edge-labeled and fully labeled graphs. In vertex-labeled graph each node has an associated label. Example of vertex-labeled graph can be a social network where each node is labeled with a person's profession. In edge-labeled graph, each graph has an associated label, often representing relationships, weights, or types. Such as in a road network, edge labels might represent distance or road type. In fully labeled graph both vertices and edges carry labels. For example a molecule represented as a graph where atoms (vertices) and bonds (edges) have labels.

A weighted graph is a type of graph in which each edge is assigned a numerical value, called a weight. These weights often represent costs, distances, capacities, or any quantifiable metric that describes the relationship between connected nodes.

A weighted graph is defined as: $G = (V, E, w)$
where:

- V is the set of vertices (or nodes),
- $E \subseteq V \times V$ is the set of edges (can be directed or undirected),
- $w : E \to R$ is a weight function assigning a real number to each edge.

Weighted graphs can be further be categorized into undirected weighted and directed weighted graphs. In undirected weighted graph, edges have no direction, i.e., if $(u, v) \epsilon E$, then $(v, u) \epsilon E$, and both share the same weight. In directed weighted graph, edges are ordered pairs; the weight of (u, v) may differ from (v, u), or the reverse edge may not exist at all. Weight of edges can be positive as well as negative. Positive weights are typical in applications like distance, time, or cost. Negative weights are allowed in some contexts (e.g., penalty, loss, or debt).

Dynamic Graphs (also known as temporal graphs or time-varying graphs) are graph structures in which the topology changes over time. These changes may include the addition or removal of nodes and edges, as well as variations in edge weights or node/edge attributes. Unlike static graphs, which represent a fixed set of relationships, dynamic graphs are better suited for modeling evolving systems such as social networks, communication networks, and transportation systems.

Let T $\subseteq R$ or Z denote the set of time steps. A dynamic graph is defined as a time-indexed sequence of graphs:

$$G = G_t = (V_t, E_t) | t \, \epsilon \, T$$

here:

- V_t: set of vertices at time t,
- $E_t \subseteq V_t \times V_t$: set of edges at time t,
- Optionally, $w_t : E_t \rightarrow R$ and ℓ_t for dynamic weights/labels.

Types of dynamics associated with dynamic graphs include node dynamics, edge dynamics and attribute dynamics. In node dynamics insertion of nodes can appear at a new time step and nodes can disappear or become inactive. In edge dynamics, edges may form or vanish over time and edge weights can evolve to reflect changes in relationship strength. With respect to attribute dynamics node or edge features (labels, weights, attributes) may vary over time. Applications of dynamic graphs include social networks, biological networks, Internet and web graphs, etc.

Types of Data Represented as Graphs

- **Social Networks:** Perhaps the most familiar application, social networks, are naturally represented as graphs. Individuals are nodes, and connections between them are edges. This representation enables the analysis of social structures, influence propagation, and community detection.
- **Transportation Networks:** Road networks, flight routes, or subway systems are examples of transportation networks. We can model transportation networks as graphs. Nodes denote locations, while edges indicate paths between them. This representation facilitates tasks like finding optimal routes, analyzing traffic flow, and designing efficient transport systems.
- **Recommendation Systems:** Graphs play a pivotal role in recommendation engines. Users and items are nodes, and interactions (e.g., purchases, likes) form the edges. By traversing the graph, these systems can suggest relevant content or products based on user behavior and preferences.
- **Biological Networks:** In biology, graphs are employed to model interactions between biological entities. For instance, in a protein-protein interaction network, proteins are nodes, and interactions are edges. This representation aids in studying cellular processes, disease pathways, and drug interactions.
- **Web Graphs:** The World Wide Web can be conceptualized as a directed graph, with web pages as nodes and hyperlinks as edges. This model underlies search engine algorithms, as it enables the computation of page rankings and relevance.

1.2.2 Representational Techniques

One common method to represent a graph is through an adjacency matrix. It is a two-dimensional array where each cell $A[i][j]$ contains information about the connection between node i and node j. In an unweighted graph, $A[i][j]$ may be 1 if there is an edge or 0 if not. In weighted graphs, the cell may contain the weight of the edge.

Alternatively, a graph can be represented using an adjacency list. In this format, each node is associated with a list of its adjacent nodes. This representation is space-efficient, especially for sparse graphs.

Another way to represent a graph is using an incidence matrix. An incidence matrix is a matrix that shows the relationship between vertices and edges. For a graph $G = (V, E)$ with n vertices and m edges, incidence matrix is an $n \times m$ matrix M, where each column represents an edge and each row represents a vertex. For undirected graphs, $M[i][j] = 1$ if vertex v_i is incident to edge e_j, 0 otherwise. For directed graphs, $M[i][j] = -1$ for the source, $M[i][j] = 1$ for the destination, 0 otherwise.

Alternatively, a graph can also be represented using an adjacency list. In this format, each node is associated with a list of its adjacent nodes. This representation is space-efficient, especially for sparse graphs.

Adjacency Matrix is ideal for dense graphs and fast edge lookups. Incidence Matrix is valuable for formal graph theory and linear algebra-based analysis. Adjacency List is the most efficient in practice for sparse and largescale graphs, especially when edge enumeration is common.

Advantages and Applications

- **Efficient Traversal:** Graph representations enable efficient traversal algorithms like depth-first search (DFS) and breadth-first search (BFS), crucial in various applications including pathfinding and network analysis.

- **Complex Relationship Analysis:** Graphs are excellent at capturing intricate relationships, allowing for the discovery of patterns and anomalies in data.

- **Scalability:** Graph databases, which store and manipulate data as graphs, are well-suited for handling large-scale interconnected datasets, making them indispensable in applications like social media analytics and recommendation systems.

- **Predictive Analytics:** By leveraging graph-based algorithms, businesses can make predictions based on patterns of interactions, leading to applications in fraud detection, product recommendations, and targeted advertising.

Representing data as graphs is a transformative approach that empowers analysts, researchers, and engineers to uncover hidden insights from complex interrelationships. As data continues to grow in size and complexity, the role of graph representations in modern information technology is supposed to become even more pivotal, revolutionizing the way we analyze and understand interconnected systems.

1.3 Pattern Recognition

Pattern recognition is a branch of machine learning that focuses on the identification and classification of patterns in data. It plays a critical role in various applications such as image processing, speech recognition, and bioinformatics, where large sets of unstructured data need to be analyzed and categorized.

The goal of pattern recognition is to assign input data to a category or class based on statistical, syntactic, or neural network approaches. It is widely used in areas such as computer vision, data mining, and artificial intelligence to automatically extract meaningful patterns from complex data.

1.3.1 Definition of Pattern Recognition

A pattern recognition system is designed to identify patterns and make decisions based on input data. At its core, pattern recognition is the process of recognizing patterns and regularities in data. The patterns can be anything from visual patterns in images, sequences of sounds, or even the arrangement of text in documents. The primary challenge lies in determining what constitutes a "pattern" and distinguishing between different patterns in noisy, real-world data. For example, in handwritten digit recognition, a system must recognize various digits written by different individuals with differing handwriting styles. In speech recognition, the system identifies phonetic patterns from speech signals and converts them into text.

1.3.2 Pattern Recognition Steps

The pattern recognition system typically follows a structured pipeline consisting of several key steps:

1. **Data Acquisition:** The first step in a pattern recognition system is collecting data from the environment. This data can come from various sources such as sensors, cameras, microphones, or other data acquisition devices.

2. **Preprocessing:** Raw data is often noisy, incomplete, or redundant. Preprocessing involves cleaning and transforming the data to improve its quality and make it suitable for analysis. Common preprocessing tasks include noise reduction, normalization, segmentation, and feature enhancement. Noise reduction removes irrelevant variations in data, such as filtering noise from an image. Normalization is done for scaling the data to a uniform range, e.g., pixel density normalization in images. Segmentation is dividing the data into meaningful parts to separate the objects from the background in an image. Feature enhancement is done for improving contrast, sharpening edges, or applying filters to highlight important information.

3. **Feature Extraction:** Identifying relevant features or characteristics from the data that help distinguish between different classes. For example, color, texture, and shape in image recognition tasks. Features should be discriminative and invariant to noise, transformations, and distortions. Types of features include structural features, statistical features, spectral features, and deep features. Structural features are shape, edges, texture, and spatial relationships. Statistical features include mean, variance, entropy, etc. Spectral features are frequency components obtained from Fourier Transform or Wavelet Transform. Deep features are high-level representations extracted using deep learning models.

4. **Classification:** Assigning the data to a predefined category or class based on the extracted features. This step involves applying classification algorithms, which could be supervised, unsupervised, or semi-supervised learning methods. Supervised learning uses labeled training data such as support vector machines, decision trees, and neural networks. Unsupervised learning identifies patterns in unlabeled data, e.g., K-Means clustering and hierarchical clustering. Semi-supervised learning uses both labeled and unlabeled data.

5. **Post-processing and Evaluation:** In some cases, the classified output undergoes further refinement, followed by performance evaluation using metrics such as accuracy, precision, recall, and F1-score. Common post-processing tasks are confidence scoring, rule-based refinement, ensemble methods, and result interpretation.

Figure 1.2 illustrates the components of a pattern recognition system. It starts with data collection by converting image, sound, or other inputs into signal data typically received from sensors. Then the collected data is preprocessed using techniques like segmentation to separate the sensed objects from the background. Data is cleaned and transformed into a suitable format. This might involve noise reduction, normalization, or feature extraction. Once the data is preprocessed, relevant features need to be extracted to represent the underlying patterns effectively. The classification step involves

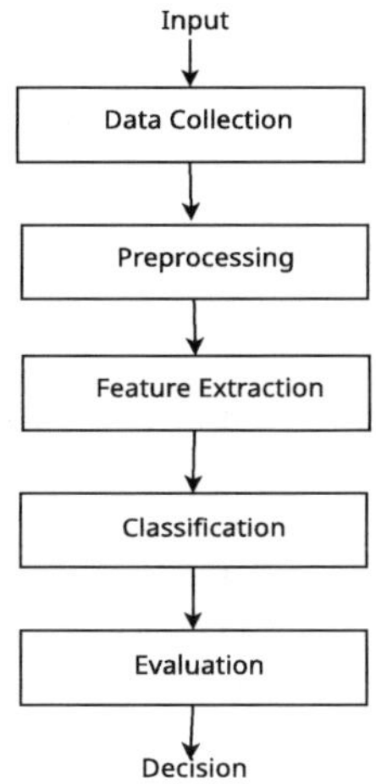

FIGURE 1.2: Pattern recognition pipeline stages

assigning the input data to a specific category or class based on extracted features. After classification, additional processing may be required to refine the results or integrate them into a decision-making system. The final step is making a decision based on recognized patterns.

A pattern recognition system follows a structured pipeline from data acquisition to decision making, ensuring that patterns are identified accurately and efficiently. The effectiveness of the system depends on proper feature extraction, selection, and classification techniques. For further refinement based on evaluation, the model needs to be adapted based on the feedback, as shown in Figure 1.3.

1.3.3 Key Approaches to Pattern Recognition

Pattern recognition approaches can broadly be divided into three categories: statistical, structural, and neural network-based.

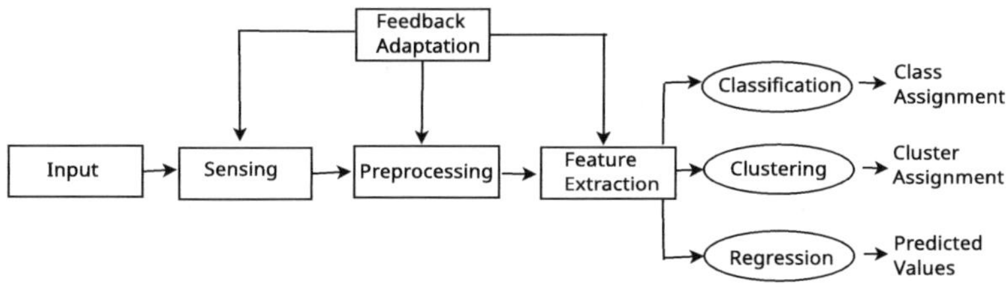

FIGURE 1.3: Pattern recognition system with adaptation

Statistical Pattern Recognition: It deals with patterns represented in numerical forms and classifies them using probabilistic and statistical models. Techniques such as Bayesian classifiers, Hidden Markov Models (HMMs), and Support Vector Machines (SVMs) fall into this category. The key idea is to model the probability distribution of each class and use statistical methods to make classification decisions.

Structural or Syntactic Pattern Recognition: This method relies on the relationships between elements in the data. Patterns are represented by formal structures such as strings, trees, or graphs. Graph matching algorithms are commonly used in this domain to find structural patterns. Structural pattern recognition is particularly useful in applications like text analysis and image interpretation, where the relationships between objects carry significant information.

Neural Network-based Pattern Recognition: Deep learning techniques, especially Convolutional Neural Networks (CNNs), have revolutionized pattern recognition, particularly in image and speech recognition. CNNs are designed to automatically learn spatial hierarchies of patterns, and they outperform traditional methods in tasks like object detection and speech-to-text conversion. Other neural network-based methods include Recurrent Neural Networks (RNNs) for time-series pattern recognition, like natural language processing.

1.3.4 Supervised and Unsupervised Learning

In pattern recognition, classification methods can be categorized based on whether they use labeled data (supervised) or unlabeled data (unsupervised).

Supervised Learning: Supervised learning involves training a model using a labeled dataset, where each example is associated with a known output or class label. Techniques like k-Nearest Neighbors (k-NN), SVM, and Neural Networks are examples of supervised learning. The system learns the patterns during the training phase and applies them to new, unseen data.

Classification is a supervised learning task in which the goal is to assign a discrete label (category) to an input based on prior training data. Objective of classification is to build a model that can map input features $x \in R^n$ to a set of predefined class labels $Y = 1, 2, ..., K$. In classification input is feature vector and output is a class label. *Regression* is a supervised learning task where the goal is to predict a continuous value from input data. Objective of regression is to learn a function $R^n \to R$ that maps input features to a real-valued output. In regression input is feature vector and output is a real number.

Unsupervised Learning: Unsupervised learning deals with unlabeled data. The goal is to find hidden patterns or groupings within the data. Clustering is an unsupervised learning task where the goal is to group similar data points into clusters without using pre-labeled data. Objective of clustering is to partition a dataset $x_1, x_2, ..., x_n$ into K groups such that data within the same cluster are more similar to each other than to those in different clusters. In clustering input is unlabeled feature vectors and output is cluster assignment. Clustering algorithms such as k-Means and Hierarchical Clustering are common unsupervised learning methods. Unsupervised learning is useful in situations where labeling data is costly or impractical.

1.3.5 Evaluation Metrics

Evaluating the performance of pattern recognition models involves various metrics, which ensure that the models are making correct classifications or predictions.

In pattern recognition, evaluation and performance metrics are essential for assessing how well a model performs on a given task—such as classification, clustering, or regression. The choice of metric depends on the nature of the problem (e.g., whether the output is categorical or continuous) and the specific goals of the system (e.g., accuracy vs. interpretability, recall vs. precision). Following are some common evaluation metrics.

Confusion Matrix is a table used to visualize the performance of a classification algorithm by showing the counts of true positive, true negative, false positive, and false negative predictions. A confusion matrix is a $K \times K$ table for a K-class classification problem showing:

- True Positives (TP): Correct positive predictions.
- True Negatives (TN): Correct negative predictions.
- False Positives (FP): Incorrectly predicted as positive.
- False Negatives (FN): Missed positive predictions.

Accuracy is the percentage of correctly classified instances over the total number of instances.

$$(TP + TN)/(TP + TN + FP + FN)$$

Precision measures how many selected items are relevant, i.e., how many predicted positives are correct.

$$TP/(TP + FP)$$

Recall measures how many relevant items are selected, i.e., how many actual positives are correctly predicted..

$$TP/(TP + FN)$$

F1-Score is harmonic mean of precision and recall.
2 ·(Precision · Recall) / (Precision + Recall)

Unlike classification tasks, clustering often lacks ground-truth labels, so evaluation focuses on how well the algorithm captures the inherent structure of the data. Clustering metrics are used to evaluate the performance and quality of clustering algorithms in unsupervised learning. Clustering metrics can be broadly categorized into internal metrics, external metrics, and relative metrics. While internal metrics require no ground truth, external metrics require ground truth labels. Internal clustering metrics evaluate clustering quality based solely on the input data and the resulting cluster assignments. They measure compactness and separation. Compactness measures how close the data points are within the same cluster. Separation measures how wellseparated the clusters are from each other. External clustering metrics compare the output of a clustering algorithm to a known ground truth. They are used in benchmarking and validation when true labels are available. Relative metrics compare clustering quality across different models or configurations.

Regression metrics are used to evaluate the performance of models that predict continuous values. These metrics quantify the error between the predicted outputs and the actual values and are crucial for model selection, tuning, and validation.

Mean Absolute Error (MAE) measures average magnitude of errors in the same units as the target variable.

Mean Squared Error (MSE) measures average of squared differences between actual and predicted values.

Root Mean Squared Error (RMSE) is similar to MAE but gives higher weight to large errors; same unit as the target variable.

1.3.6 Applications of Pattern Recognition

Pattern recognition is integral to numerous applications across diverse fields.

Image Recognition: Pattern recognition is widely used in image processing, particularly in facial recognition, object detection, and medical imaging. CNNs have achieved significant success in this domain by learning hierarchical patterns in images.

Speech Recognition: In speech-to-text systems, pattern recognition helps convert spoken language into written text. Deep learning models like RNNs and Long Short-Term Memory (LSTM) networks are commonly employed for this purpose.

Bioinformatics: Pattern recognition plays a crucial role in analyzing biological data such as gene sequences, protein structures, and drug discovery. Sequence alignment and motif recognition are key techniques used in bioinformatics.

Financial Markets: In finance, pattern recognition is used to identify trends and make predictions based on historical stock market data. Time-series analysis, supported by machine learning models, helps analysts make more accurate forecasts.

1.3.7 Challenges in Pattern Recognition

Despite the advances, several challenges persist in pattern recognition.

- **Variability in Data:** Real-world data is often noisy, incomplete, or contains variations that make pattern recognition difficult.
- **Scalability:** Handling large datasets requires robust and scalable algorithms that maintain performance under massive data loads.
- **Interpretability:** As models become more complex, especially with deep learning, understanding and interpreting the learned patterns becomes challenging.
- **Generalization:** Ensuring that a model trained on a specific dataset generalizes well to new, unseen data remains a central challenge in machine learning and pattern recognition.

1.3.8 Future Directions

The future of pattern recognition is likely to be shaped by further advancements in deep learning, reinforcement learning, and unsupervised learning. The integration of artificial intelligence with pattern recognition opens up new possibilities in areas like autonomous driving, healthcare diagnostics, and smart cities. Moreover, the increasing availability of large-scale datasets will drive further innovations in this field.

Terms artificial intelligence, pattern recognition, machine learning, and deep learning are interrelated. The relationship between these concepts can be understood as a hierarchical structure, where deep learning is a subset of pattern recognition and machine learning, and machine learning is a subset of artificial intelligence.

Artificial Intelligence (AI) is a broad field of computer science that focuses on creating systems that can simulate human intelligence. AI encompasses a wide range of techniques, including rule-based systems, search algorithms, knowledge representation, expert systems, and learning-based approaches. Machine Learning (ML) is a subset of AI that enables systems to learn from data and make predictions or decisions without explicit programming. Instead of using predefined rules, ML systems identify patterns from data and adjust their models accordingly. Deep Learning (DL) is a subset of Machine Learning that utilizes artificial neural networks with multiple layers (deep architectures) to learn complex patterns in large datasets. It is inspired by the structure of the human brain.

The integration of AI with pattern recognition opens up new possibilities in areas like autonomous driving, healthcare diagnostics, and smart cities. Moreover, the increasing availability of large-scale datasets will drive further innovations in this field. AI is the umbrella term encompassing various approaches to building intelligent systems. Pattern Recognition is a subfield of AI focused on detecting and classifying patterns. ML is a powerful tool for pattern recognition, enabling computers to learn from data rather than relying on hand-crafted rules. DL is an advanced ML technique that has revolutionized pattern recognition by automatically learning high-level features from data. As technology advances, these fields continue to evolve and overlap, leading to more powerful AI systems capable of solving complex real-world problems.

1.4 Organization of the Book

This book is organized as follows. Chapter 2 describes an overview of structural pattern recognition. This chapter delves into the theory, techniques, and applications of structural pattern recognition, exploring how structured data representations such as graphs and trees can be used to recognize complex patterns.

Chapter 3 provides a comprehensive review of various techniques and algorithms used in exact, approximate, and error-tolerant graph matching. The process of graph matching involves computing the similarity between two graphs, and it can be broadly classified into two categories: exact and inexact graph matching. Exact graph matching has more theoretical implications in computer science, whereas inexact graph matching bears greater practical significance in computer science and pattern recognition.

Chapter 4 describes extensions of graph edit distance, which offer a trade-off between computation time and accuracy in various graph matching applications. It introduces an approach to error-tolerant graph matching utilizing graph homeomorphism, enabling the computation of structural similarity between two graphs. In this technique, path contraction plays a pivotal role in reducing the number of nodes in the input graphs while preserving their topological equivalence. Additionally, this chapter presents an alternative method for error- tolerant graph matching employing node contraction. This technique transforms a given graph into another by contracting smaller-degree nodes. These approaches effectively reduce the search space required to calculate graph edit distances between transformed graphs.

Graph matching using centrality measures is presented in Chapter 5. The applications of exact graph matching in real-world scenarios are often limited due to the presence of noise or errors during graph processing. As an alternative, error-tolerant graph matching provides a solution for performing approximate graph matching. However, the exponential complexity associated with graph edit distance prompts the exploration of other methods that offer more efficient graph matching, albeit with a slight decrease in accuracy. To achieve size reduction in graphs, depending on the structure and properties of the dataset, we can employ different centrality measures to identify and remove nodes. In this chapter, we utilize centrality measures such as eigenvector, betweenness, and PageRank, in addition to degree centrality, to reduce the size of the graphs and estimate an early approximate graph matching between two graphs. These approaches help strike a balance between computational efficiency and matching accuracy, making them suitable choices for specific graph matching tasks.

Chapter 6 presents an approach to error-tolerant graph matching using geometric graphs. Geometric graph matching involves assessing the similarity between geometric graphs, where each vertex is associated with a coordinate point in a plane. Exact matching aims to establish a precise correspondence between two geometric graphs, while error-tolerant matching calculates an approximate similarity between them. It introduces the concepts of vertex distance or dissimilarity and edge distance or dissimilarity between two geometric graphs. We combine vertex distance and edge distance to computer Graph Distance, which enables exact and error-tolerant geometric graph matching. This approach facilitates efficient and accurate comparison between geometric graphs, allowing for practical applications in various fields.

Chapter 7 describes the novel concepts of graph kernel and embedding to perform graph matching. The graph kernel provides us with a way to embed the space of graphs into an inner product space, utilizing the graph kernel similarity measure. One can enhance graph matching techniques by adopting ideas originally designed for statistical pattern recognition represented by feature vectors to structural pattern recognition. To tackle the graph matching problem within a vector space connected to the space of graphs, we only need to define a similarity measure for graphs that meets certain properties. It enables us to extend powerful pattern recognition concepts to the domain of graphs and enhance our graph-matching capabilities.

Chapter 8 explores the fundamental concepts, techniques, and applications of graph matching in image analysis. Image analysis is a crucial field within computer vision, aiming to extract meaningful information from images. Graph matching is a powerful technique employed in image analysis to compare and align structural representations of objects or scenes represented as graphs.

Chapter 9 explores the fundamental concepts, methodologies, and applications of graph matching in Social Network Analysis. Social Network Analysis (SNA) is a powerful tool for understanding the complex relationships and interactions among individuals, groups, or entities within a social system. Graph matching techniques play a significant role in SNA, assisting in the comparison and alignment of social networks to uncover patterns, detect communities, and identify structural similarities.

Finally, Chapter 10 describes the recent advances, future directions, and emerging trends in structural pattern recognition and graph matching.

Chapter 2

Structural Pattern Recognition

Patterns are ubiquitous in the world, from the symmetrical petals of a flower to the repeating melodies in music. In various fields, recognizing and interpreting these patterns is fundamental to understanding, analyzing, and making informed decisions. For the domain of computer science and artificial intelligence, the study of structural pattern recognition stands at the forefront of this endeavor.

Structural Pattern Recognition (SPR) is a branch of pattern recognition that focuses on the relationships and structures within data, often represented by graphs or other structured models. Unlike statistical pattern recognition, which treats data as a set of independent features, SPR captures complex relationships among features, making it particularly useful for problems where the organization or configuration of parts is critical, such as image analysis, natural language processing, and bioinformatics.

This chapter delves into the theory, techniques, and applications of structural pattern recognition, exploring how structured data representations such as graphs and trees can be used to recognize complex patterns. This chapter is organized as follows. Section 2.1 presents an introduction to structural pattern recognition. Section 2.2 describes the basic concepts in structural pattern recognition and its differences with statistical pattern recognition. Section 2.3 introduces the structural representations, including string-based, tree-based, graph-based and formal grammar-based representations. Section 2.4 describes the algorithms for structural pattern recognition. Learning in structural pattern recognition is depicted in Section 2.5. Important applications of structural pattern recognition is listed in Section 2.6, and finally Section 2.7 describes the challenges in structural pattern recognition.

2.1 Introduction to Structural Pattern Recognition

Structural pattern recognition is based on the premise that many real-world objects and phenomena are best represented not as isolated sets of features but as structured entities, where the relationships between components play an essential role in defining their identity.

Pattern recognition can be broadly classified into two categories: statistical pattern recognition and structural pattern recognition, depending on the

DOI: 10.1201/9781003532248-2

nature of the underlying problem. Statistical Pattern Recognition relies on the use of feature vectors to encapsulate and represent distinct patterns. This methodology utilizes the well-established mathematical techniques derived-from vector space theory, rendering them directly applicable in the domain of statistical pattern recognition. This leads to the development of a host of efficient algorithms, enhancing the accuracy and effectiveness of pattern recognition systems. However, a significant limitation arises when dealing with patterns characterized by intricate structures like trees or graphs. In such scenarios, the conventional use of feature vectors with fixed dimensions proves inadequate. These complex structures cannot be adequately captured within the constraints of fixed-dimensional vectors. In contrast, Structural Pattern Recognition offers an alternative by recognizing patterns through graph-based representations instead of relying on feature vectors, as illustrated in Figure 2.1. The notable advantage of utilizing graphs lies in their higher representational capacity compared to feature vectors. This introduces a classic trade-off between the use of vectors and graphs. While vectors achieve computational efficiency, they lack representational power. Conversely, graphs exhibit robust representational abilities, although at the cost of reduced computational efficiency.

	Representational Power	Computational Efficiency
Structural Pattern Recognition	High	Low
Statistical Pattern Recognition	Low	High

FIGURE 2.1: Structural versus statistical pattern recognition

Structural Pattern Recognition, often abbreviated as SPR, is a branch of machine learning and artificial intelligence that focuses on recognizing patterns in complex, structured data. At the heart of SPR lies the concept of structure. In data, structure refers to the organization, arrangement, or interconnection of elements. It can manifest in various forms, such as trees, graphs, sequences, or more intricate structures. SPR aims to develop algorithms and techniques to distinguish these underlying structures, allowing for a deeper understanding and interpretation of the data.

2.2 Key Concepts in Structural Pattern Recognition

SPR is a subfield of pattern recognition that focuses on analyzing and recognizing complex data structures by representing them as graphs, trees, or other relational models. Unlike statistical or feature-based pattern recognition, which relies on numerical feature vectors, structural pattern recognition emphasizes the relationships between components in structured data. Figure 2.2 describes the stages of the SPR System. Recognition components consist of preprocessing, pattern representation, and structural analysis. Pattern representation itself consists of feature extraction, segmentation, and primitive recognition. The analysis component for the sample patterns consists of primitive selection and structural inference.

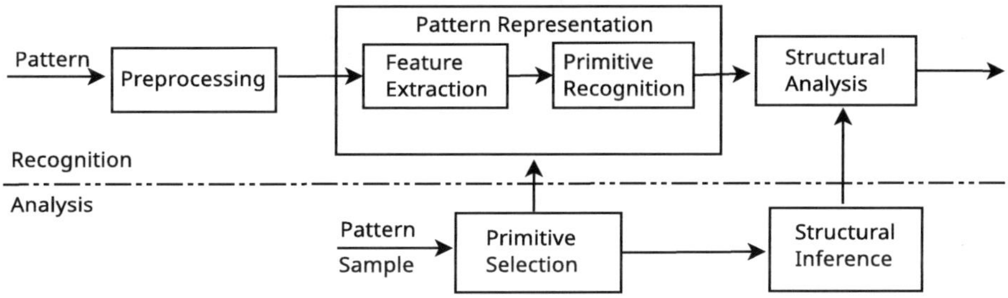

FIGURE 2.2: Stages of structural pattern recognition system

- **Patterns:** In SPR, patterns are often represented as structures such as graphs, trees, or grammars, where components (e.g., nodes, symbols) are connected by relationships (e.g., edges, rules).
- **Representation:** A primary task in SPR is to find appropriate representations for structured data. Graphs, strings, and hierarchical structures are common representations.
- **Matching:** Matching patterns in SPR involves finding correspondences between structured objects, often requiring techniques such as graph matching, string matching, or tree matching.

Objects or patterns are modeled as graphs where nodes represent components (e.g., image segments, words in text, atoms in molecules) and edges define relationships (e.g., spatial proximity, syntactic rules, chemical bonds). Some structured data, such as natural language or hierarchical objects, are represented using trees where each node has a parent-child relationship. Patterns can also be represented as sequences, such as DNA sequences in bioinformatics or time-series data in speech recognition.

2.2.1 Feature Extraction in Structural Data

Feature extraction is a crucial step in SPR, where patterns are represented as structured entities such as graphs, trees, strings, or other relational models. Unlike statistical pattern recognition, which relies on numerical feature vectors, structural pattern recognition focuses on the relationships between elements in a pattern, making feature extraction more complex but also more expressive.

Feature extraction in structural pattern recognition refers to the process of deriving meaningful representations from structured data while preserving the relationships between components. The goal is to convert raw structural patterns into a form that can be efficiently processed for classification, clustering, or other machine learning tasks. Instead of simple numerical attributes, structural patterns often contain topological, geometric, or relational features, such as:

- String or sequence-based features (e.g., substrings, edit distances)
- Tree-based features (e.g., subtree structures, hierarchical relationships)
- Graph-based features (e.g., nodes, edges, connectivity)

Sequence-Based Feature Extraction

Patterns in sequential data, such as DNA sequences, speech signals, or time-series data, are often represented as strings or sequences. Feature extraction techniques include: edit distance, hidden Markov model, and n-gram model. Edit distance features measure the similarity between two sequences based on operations like insertion, deletion, and substitution of symbols. Hidden Markov model captures probabilistic dependencies in sequential patterns. An n-gram model extracts fixed-length subsequences to capture local dependencies.

Tree-Based Feature Extraction

Hierarchical structures such as syntax trees, XML data, and organizational charts can be represented as trees. Feature extraction from trees often involves subtree patterns for identifying frequently occurring subtree structures. Tree kernel methods are used to measure the similarity between trees using recursive decomposition. Path features can be applied for extracting sequences of nodes or edges along specific paths. For example, in natural language processing, sentences can be parsed into syntactic trees, where the extracted features help in tasks like sentiment analysis or text classification.

Graph-Based Feature Extraction

Many real-world patterns, such as molecules, social networks, and handwriting strokes, can be naturally represented as graphs. Feature extraction in graphs involves deriving numerical or symbolic attributes from their structure. Graph embedding is one of the important techniques of feature extraction by converting graphs into vector representations using techniques such as graph kernels, graph neural networks, and spectral embedding. Graph embedding techniques are methods that map graph-structured data into a lower-dimensional vector space while preserving the graph's structural and semantic properties. These embeddings transform nodes, edges, or entire subgraphs into fixed-length numerical vectors, enabling the application of machine learning and data analysis techniques that work with vector-based data.

Structural descriptors are features that summarize structural properties, such as node degree distribution, betweenness centrality, or eigenvector centrality. Subgraph-based features identify frequently occurring subgraphs or motifs within larger graphs. Graph spectral features are captured using eigenvalues and eigenvectors of the graph Laplacian matrix. ***Graph descriptors*** can be used to extract features such as node degree, clustering coefficient, shortest path lengths, and centrality measures. Graph descriptors are quantitative or qualitative measures that capture structural properties of a graph. They are used in various applications, including pattern recognition, machine learning, bioinformatics, and social network analysis, to characterize graphs and compare their structures efficiently. Graph descriptors help in transforming complex graph structures into compact numerical representations that can be used for classification, clustering, and similarity analysis. These descriptors preserve essential information about the nodes, edges, connectivity, and topology of a graph.

Graph descriptors can be categorized into global, local, and spectral descriptors based on the level of structural information they capture. ***Global descriptors*** summarize the overall structure of a graph. They provide insight into graph size, connectivity, and complexity.

Number of nodes ($|V|$) and edges ($|E|$) are basic descriptors representing the size of a graph.

Graph density measures how connected a graph is. Density for an undirected graph is defined by: $Density = \frac{2|E|}{|V|(|V|-1)}$.

Average node degree is the mean number of edges per node indicating overall connectivity.

Graph diameter is the length of the shortest path between the most distanced node.

Clustering coefficient measures the likelihood that two neighbors of a node are also connected, indicating graph transitivity.

Graph entropy quantifies graph complexity based on information-theoretic measures.

Local descriptors focus on individual nodes and their neighborhoods. ***Degree*** of a node is equal to the number of edges connected to a node.

Eccentricity is the maximum distance from a node to any other node in the graph. ***Centrality measures*** are indicators of a node's importance in a graph. These include degree centrality, betweenness centrality, closeness centrality, and eigenvector centrality.

Spectral graph descriptors are derived from the eigenvalues and eigenvectors of matrices associated with a graph, such as the adjacency matrix and Laplacian matrix. The eigenvalue spectrum provides a fingerprint of a graph structure. Graph Laplacian eigenvalues capture connectivity and diffusion properties. Spectral radius is the largest eigenvalue of the adjacency matrix related to graph stability.

Graph descriptors provide a compact and informative representation of graphs, enabling efficient analysis and comparison. By extracting global, local, and spectral properties, they play a vital role in graph-based machine learning, pattern recognition, and network science.

Feature extraction in structural pattern recognition involves transforming complex relational patterns into a format suitable for machine learning. Techniques such as graph-based embeddings, tree-based kernels, and sequence models enable effective pattern analysis across diverse domains, including image recognition, NLP, bioinformatics, and social network analysis. While feature extraction for structured data is more complex than for statistical patterns, it provides richer representations that can significantly improve recognition and classification performance.

2.2.2 Difference Between Statistical and Structural Pattern Recognition

Pattern recognition techniques can be broadly classified into Statistical Pattern Recognition and Structural Pattern Recognition. Both approaches aim to analyze and classify patterns, but they differ in how they represent data, extract features, and perform classification. A summary of comparisons between statistical and structural pattern recognition is provided in Table 2.1.

TABLE 2.1: Statistical vs. Structural Pattern Recognition

Aspect	**Statistical Pattern Recognition**	**Structural Pattern Recognition**
Representation	Vectors in a multi-dimensional space	Structured objects, graphs, trees
Feature Type	Numeric features	Structural and relational
Pattern Classification	Distance measure, prob. dist.	Structural similarity
Data Nature	Suitable for patterns in vector form	Relational, topological structures
Common Techniques	SVM, PCA, FDA	Graph Matching, GED, TED, SED
Complexity	Computationally efficient	Computationally expensive
Robustness	Less robust to structural variations	More robust to variations in pattern

2.2.2.1 Statistical Pattern Recognition

Statistical pattern recognition is a fundamental approach in machine learning and pattern analysis that aims to classify patterns based on extracted features and an underlying statistical model that describes the generation process of these patterns. It focuses on feature-based representations, where objects are described by vectors of independent features, and recognition tasks involve classification based on these features. The general procedure for statistical pattern recognition follows a structured framework of feature extraction, model training followed by pattern classification. *Feature extraction* identifies and extracts meaningful characteristics from raw data, forming a feature vector representation. *Model training* uses a set of labeled training data to train a classifier based on statistical principles. *Pattern classification* applies the trained model to new, unseen patterns to assign them to appropriate categories. This methodology has proven highly effective in many applications, such as image recognition, speech processing, and bioinformatics, where patterns can be well-represented using numerical feature vectors.

While statistical pattern recognition is a powerful tool, not all pattern recognition problems can be easily addressed using feature vectors. Many real-world patterns contain structural and relational information that is difficult or even impossible to quantify in a conventional feature-based representation. Some key challenges include:

- **Loss of structural information**: Representing a pattern as a simple feature vector often disregards important spatial, topological, or sequential relationships within the data. For instance, in handwritten character recognition, a feature vector may capture stroke widths and angles but may fail to encode how different strokes are connected.
- **Complex and high-dimensional relationship**: Many patterns, such as graphs, trees, and sequences, exhibit relationships that cannot be efficiently captured using fixed-length feature vectors. Such as, in molecular biology, protein structures are often best represented as graphs, where nodes correspond to atoms and edges to chemical bonds. A feature vector may fail to capture such intricate relationships effectively.
- **Variability and deformation in patterns**: Some pattern recognition tasks involve high intra-class variation, where objects belonging to the same category exhibit significant differences. Face recognition, for example, must account for variations in lighting, pose, and expression, making it difficult to capture a meaningful feature representation.
- **Data representation limitations**: In domains such as natural language processing, patterns are inherently sequential and context-dependent. A simple numerical vector representation of a sentence may ignore word order, syntax, and semantics, leading to suboptimal classification results.

To address these limitations, structural pattern recognition techniques focus on alternative representations that preserve relational and structural information. Statistical pattern recognition is a well-established approach that has demonstrated great success in many applications. However, its reliance on feature vector representations poses challenges when dealing with structured or relational data. As a result, researchers and practitioners increasingly explore graph-based models, kernel methods, and deep learning approaches to overcome these limitations. By leveraging these advanced techniques, pattern recognition systems can achieve greater accuracy and robustness in complex real-world scenarios.

2.3 Structural Representations in Pattern Recognition

In SPR, data is often represented as structured objects such as graphs, trees, strings, or formal grammars. These representations are particularly suited for capturing relational patterns.

2.3.1 String-Based Representations

String-based representation is a method used in pattern recognition where objects, patterns, or signals are expressed as sequences of symbols (or characters) from a predefined alphabet. This approach is particularly useful for structured data, where the relationships between elements are important and need to be preserved. String representations are often used when patterns consist of sequences of symbols, such as DNA sequences, text, or time series.

A string is an ordered sequence of symbols chosen from a finite set, known as an alphabet. Objects are mapped to strings that encode structural, spatial, or sequential information. Applications of string-based representation range from text and speech recognition, biometric identification, and bioinformatics to network intrusion detection. Text documents and spoken words are often analyzed using string representations to recognize words, phrases, or entire sentences. For example, Optical Character Recognition (OCR) uses string-based methods to convert scanned images of text into editable text. In bioinformatics, DNA and protein sequences are naturally represented as strings over a limited alphabet (e.g., A, T, C, G for DNA). Structural pattern recognition techniques also help in identifying mutations, evolutionary relationships, and genetic markers. Strokes in handwriting or movements in gestures can be encoded as sequences of symbols. A signature verification system can analyze variations in handwritten signatures using string-based matching techniques. Features such as fingerprints or iris scans can be represented as sequences of extracted features, allowing for efficient matching and recognition. String matching techniques are used to identify suspicious patterns in network traffic by comparing packet sequences with known attack signatures.

Primary advantages of string-based representation are that it captures sequential information and it performs efficient matching and comparison. Since strings are sequential, they retains the order of elements, making them suitable for temporal and structural pattern recognition. Because of the simplicity of its presentation, many well-defined algorithm exist for comparing and searching strings. Due to its efficiency, it is used in fields ranging from text analysis to biological data processing. Limitations of string-based representation include loss of geometric and spatial information and sensitivity to noise. As strings may not fully capture spatial relationships, making them unsuitable for certain tasks (e.g., image recognition). It is also sensitive to noise, as small changes in data (e.g., insertion or deletion of a character) can significantly alter the string representation, making recognition more challenging.

String-based representation is a powerful approach in pattern recognition, especially for problems involving sequential or structured data. It is widely used in text processing, bioinformatics, speech recognition, and biometric applications. However, its effectiveness depends on the choice of encoding method, similarity measure, and computational efficiency.

2.3.2 Tree-Based Representations

Tree-based representations are hierarchical data structures used in pattern recognition to efficiently model, analyze, and classify structured data. Trees are hierarchical structures where each node (except the root) has a single parent, and they are widely used in domains where hierarchical relationships are important. These representations organize information, where nodes represent data elements, and edges define relationships between them. Tree-based models are particularly useful for handling structured, hierarchical, or relational data, such as text, images, biological sequences, and graphs. Following are some of the key features of tree-based representations.

1. **Hierarchical Structure:** Data is represented in a multi-level format, capturing different levels of abstraction. Root nodes define the global structure, while leaf nodes represent finer details.

2. **Recursive Nature:** Trees naturally encode recursive structures, making them ideal for representing objects composed of smaller parts (e.g., syntactic structures in language, segmentation in images).

3. **Efficient Search and Retrieval:** Many tree-based structures allow fast searching, sorting, and pattern matching. Examples include decision trees, suffix trees, and KD-trees.

4. **Flexible Representation for Complex Data:** Trees can represent diverse data types, such as sequences (e.g., parse trees in NLP),

hierarchical clustering (e.g., dendrograms), and structured objects (e.g., scene graphs in computer vision).

Depending on the application domain different types of tree-based representations are applied in pattern recognition. Following are some of the important types of tree-based representations.

1. **Decision Trees:** In machine learning, decision trees are used to represent decision rules that partition the data. Each internal node represents a decision based on a feature, and leaves correspond to class labels. Used in classification and regression tasks (e.g., CART, ID3, C4.5). As an example, a decision tree can be used for email classification to classify spam and non-spam emails.

2. **Parse Trees:** In natural language processing, parse trees or syntax trees represent the syntactic structure of sentences. Example includes, a parse tree representing the grammatical structure of a sentence.

3. **Suffix Trees:** It is a compressed tree containing all the suffixes of the given text. Suffix trees provides a particularly fast implementation for many important string operations. It efficiently stores substrings for fast searching and matching. It is commonly used in string processing and bioinformatics such as DNA sequence analysis.

4. **KD-Trees:** It is a binary search tree where data in each node is K-dimensional point in space. It is used for spatial data partitioning to organize points in a K-dimensional space. KD-trees help in fast nearest-neighbor queries in high-dimensional spaces.

5. **Region Adjacency Trees (RATs):** Applied in image segmentation, where nodes represent segmented regions and edges define adjacency relationships.

Tree-based representations play a crucial role in pattern recognition by providing efficient hierarchical models for structured data. Whether used in classification, segmentation, or clustering, trees enable better data organization, efficient computation, and improved interpretability in various applications across computer vision, NLP, bioinformatics, and machine learning.

2.3.3 Graph-Based Representations

Graphs are one of the most commonly used representations in SPR due to their ability to model relationships between entities. Graph-based representations in pattern recognition refer to the use of graphs as a structured way to model and analyze complex patterns, relationships, and dependencies in data. Unlike traditional vector-based representations, graphs provide a flexible and expressive framework for capturing structural and relational information in a variety of applications, including image analysis, bioinformatics, social network analysis, and natural language processing.

A *graph* is a mathematical structure represented as $G = (V, E)$, where V is a set of nodes (or vertices) that represent entities in the pattern recognition task. $E \subseteq V \times V$ is a set of edges that define relationships between nodes. *Attributed Graphs* are graphs in which nodes and edges have attributes to provide additional semantic meaning. In real-world applications, nodes and edges often have attributes (e.g., labels, weights), resulting in attributed graphs that are more expressive. In *Subgraph representation* objects or patterns can often be represented as subgraphs of a larger graph, making subgraph matching a crucial task in SPR. Graphs are sometime categorized as a directed and undirected graphs depending on whether the edges have directions or not.

As discussed above there are several advantages of graph-based representations. Graphs are rich in structural representation and they naturally model hierarchical, sequential, and spatial relationships between objects. Graph are flexible and scalable which can represent irregular structures such as molecules, social networks, and handwritten characters. Graph-based methods can effectively capture invariance in transformations such as rotation, translation, and scaling. Unlike traditional feature vectors, graphs can represent data that does not fit into a regular Euclidean space. Graph-based representations are a powerful approach in pattern recognition, providing structural, relational, and contextual insights that traditional vector-based methods may overlook. With advancements in graph neural networks and graph matching algorithms, graph-based techniques continue to evolve, playing a crucial role in modern pattern analysis tasks.

2.3.4 Formal Grammar-Based Representations

Formal grammar-based representation is a method used in pattern recognition where patterns are represented and analyzed using formal grammars. Grammars are used to describe the structure of patterns, where a set of production rules defines how patterns can be generated. This approach is particularly useful for recognizing structured patterns, such as natural language processing, biological sequences, and syntactical pattern analysis.

A ***formal grammar*** is a set of rules that define a language. It consists of a set of symbols and production rules that describe how strings in a language can be generated or parsed. In pattern recognition, formal grammars help model patterns and extract meaningful structures from raw data.

Definition 2.1 (Formal Grammar) A *formal grammar* G is typically defined as:

$$G = (V, \Sigma, P, S)$$

where:

- V : A finite set of non-terminal symbols

- Σ : A finite set of terminal symbols
- P : A finite set of production rules that define how symbols can be replaced.
- $S \in V$: The starting symbol from which all valid strings are derived. □

Following is a simple example of context-free grammar for arithmetic expression over alphabet set $\{a, b\}$.

$$S \rightarrow S + S | S * S | (S) | a | b$$

This grammar generates all valid arithmetic expression with addition (+) and multiplication (∗) over symbols a and b.

Formal grammar-based representation are used for defining rules for pattern recognition (e.g., DNA sequence modeling), recognizing patterns with hierarchical structures such as parsing sentences in NLP. It is also used in handling complex dependencies between pattern elements and improving interpretability and consistency in structured data processing.

In handwritten digit recognition, a grammar-based approach can define how strokes are combined to form digits. A simple grammar might describe the starting point of a digit, the order of strokes to complete a number, and variations in stroke direction that still produce the same digit. Instead of treating digit images as raw pixel data, a grammar-based approach models their structural composition, making recognition more robust to noise and variations.

Formal grammars are categorized based on their expressive power, following the Chomsky classification of languages.

Regular grammar also known as type 3 grammar uses simple rules where each production has at most one non-terminal followed by a terminal. These are equivalent to *finite automata.* This model is efficient but lacks the ability to capture complex hierarchical structure. Example of production rules include: $A \rightarrow aB, B \rightarrow b$. Regular grammar is applicable in simple text pattern matching such as regular expressions in search engines.

Context-Free Grammar (type 2) allows rules where a single non-terminal can produce a sequence of symbols including terminals and non-terminals. Captures nested structures, such as parenthesis matching or sentence parsing. For example the grammar $S \rightarrow aSb|\varepsilon$ generates strings like "ab", "aabb" etc. Applications includes speech recognition, biological sequence analysis, and programming language parsing. Compilers use context-free grammar to process code syntax.

Context-Sensitive Grammar (type 1) can have multiple symbols including terminals and non-terminals on both left-hand and right-hand of production rule. Since it can have context in both left and right hand side of production rule, it is called as context-sensitive grammar. Production rules $AB \rightarrow CD, aB \rightarrow Cd$ belongs to context-sensitive grammar.

TABLE 2.2: Applications of Grammar-based Representation

Application Area	Grammar Type	Use Case
Speech Recognition	Context-Free	Parsing spoken sentences
Handwriting Recognition	Context-Free	Modeling character structure
Bioinformatics	Context-Sensitive	DNA and protein sequence alignment
OCR	Regular	Recognizing printed text patterns
Computer Vision	Context-Free	Recognizing objects in structured scenes
NLP	Context-Sensitive	Understanding complex sentence structures

Unrestricted Grammar (type 0) is the most powerful type, capable of describing Turing machine languages. It is used in theoretical computing, mathematical modeling of neural network and advanced AI language models. A summary of applications of various grammar-based representation in patter recognition is provided in Table 2.2.

Stochastic Grammars extend classical grammars with probabilistic rules, allowing for the generation of patterns with uncertainty. Stochastic grammar also known as probabilistic grammar is a type of formal grammar where production rules are assigned probabilities. These probabilities represent the likelihood of a particular rule being used when generating or parsing a string. Stochastic grammars are widely used in natural language processing, speech recognition, bioinformatics, and machine learning to model uncertain or probabilistic structures in sequences.

Grammar-based representation captures hierarchical and recursive structures. Rules can be explicitly defined, making patterns understandable. It can handle variations in handwriting, speech, and bio-sequences. Grammar-based representations in pattern recognition provide structured, interpretable, and hierarchical ways to analyze patterns. From speech recognition to biological sequence analysis, formal grammars help in defining the underlying structure of patterns, making recognition tasks more efficient and explainable. However, choosing the right type of grammar depends on the complexity and nature of the pattern recognition problem.

2.4 Techniques for Structural Pattern Recognition

Several techniques have been developed for recognizing patterns in structured data, ranging from exact matching algorithms to machine learning-based approaches. Matching is the heart of SPR, as it allows the comparison of different structures to recognize pattern.

2.4.1 String Matching

String matching is the process of finding occurrences of a pattern (a smaller string) within a text (a larger string). String matching algorithms are fundamental techniques in computer science used to find occurrences of a pattern within a text. These algorithms have widespread applications in text searching, data mining, bioinformatics, and information retrieval.

Assume that text is an array $T[1...n]$ of n characters, and the pattern $P[1...m]$ is an array of m characters, where $m \leq n$. The pattern P occurs with shift s in text T, when $0 \leq s \leq n-m$ and $T[s+1...s+m] = P[1...m]$ so that $T[s+j] = P[j]$, for $1 \leq j \leq m$. When pattern P occurs with shift s in text T, then s is called as a valid shift. Given a pattern P and a text T, ***string-matching problem*** is defined as a problem of finding all valid shifts of pattern P in the text T.

The Simple-String-Matching algorithm searches all shifts using a *for* loop which checks the condition $P[1...m] == T[s+1...s+m]$ for every $n-m+1$ possible values of s. Observe that for each of the $n-m+1$ values of the shift s, the loop in line 4 executed m times to verify the shift. The worst case running time of Algorithm 1 is $\Theta((n-m+1)m)$.

Knuth-Morris-Pratt Algorithm

The Knuth-Morris-Pratt (KMP) algorithm is an efficient string-matching algorithm designed to find occurrences of a pattern P of length m with a text T of length n. Unlike the simple brute force approach, which can lead to redundant comparisons, KMP avoids unnecessary re-checks by preprocessing the pattern using a prefix function (also called the partial match table). KMP-String-Matching is described in Algorithm 2. The KMP-String-Matching algorithm calls the procedure Compute-Prefix-Function to compute f in line 3. The prefix function for a given pattern encodes information about the way the pattern aligns with its own shifts. This knowledge can be leveraged to optimize pattern-matching algorithms by eliminating redundant or unnecessary shift evaluations, thereby improving efficiency compared to the naive approach

Algorithm 1 : Simple-String-Matching (T, P)

Input: A text T and pattern P
Output: Position of pattern P in text T

```
1: n ← length[T]
2: m ← length[P]
3: for (s ← 0 to n − m  do
4:     if (P[1...m] == T[s + 1...s + m]) then
5:         print Pattern exists with shift s
6:     end if
7: end for
```

Algorithm 2 : KMP-String-Matching (T, P)

```
Input: A text T and pattern P
Output: Position of pattern P in text T
1: n ← length[T]
2: m ← length[P]
3: f = Compute-Prefix-Function(P)
4: q = 0
5: for i ← 1 to n do
6:     while (q > 0 and P[q + 1] ≠ T[i] do
7:         q = f[q]
8:         if (P[q + 1] == T[i]) then
9:             q ← q + 1
10:        end if
11:        if (q == m) then
12:            print Pattern exists with shift i − m
13:            q ← f[q]
14:        end if
15:    end while
16: end for
17: procedure Compute-Prefix-Function(P)
18:    m ← length[P]
19:    let f[1...n] be an array
20:    f[1] ← 0
21:    k ← 0
22:    for q ← 2 to m do
23:        while (q > 0 and P[q + 1] ≠ P[q] do
24:            k ← f[k]
25:            if (P[q + 1] == P[q]) then
26:                q ← q + 1
27:            end if
28:            f[q] ← k
29:        end while
30:    end for
31:    return f
32: end procedure
```

in Algorithm 1. The variable q in line 4 keep track of number of characters matched. The *for* loop in line 5 scan the text from left to right to match the subsequent characters.

The running time of the KMP-String-Matching algorithm is $\Theta(n)$, whereas the running time of Compute-Prefix-Function is $\Theta(m)$. The Simple-String-Matching algorithm as well as the KMP-String-Matching algorithm are examples of exact matching between two strings, where we want to search for a pattern P in a text T. Now we describe the inexact technique of string matching. The most important one is the edit distance between two strings, known as string edit distance which is equal to the minimum number of insertions, deletions, and substitutions of characters needed to transform the input string to the output string.

String Edit Distance

String edit distance also known as Levenshtein distance is defined as the minimum number modifications in terms of additions, deletions and substitutions of characters or symbols needed to transform one string into another.

Definition 2.2 (String Edit Distance) *String edit distance* between two strings A and B is the minimum number of edit operations required to transform string A to string B, where edit operations are addition, deletion and substitution of characters in the string.

Algorithm 3 : String-Edit-Distance (A, B)

Input: Two strings A of length m and B of length n
Output: String edit distance between A and B

```
1: m ← length[A]
2: n ← length[B]
3: for (i ← 0 to m) do
4:     d[i][0] ← i
5: end for
6: for (j ← 0 to n) do
7:     d[0][j] ← j
8: end for
9: for (i ← 1 to m) do
10:     for (j ← 1 to n) do
11:         if (A[i − 1] == B[j − 1]) then
12:             d[i][j] ← d[i − 1][j − 1]
13:         else d[i][j] ← min(d[i][j − 1] + 1, d[i − 1][j] + 1, d[i − 1][j − 1] + 1)
14:         end if
15:     end for
16: end for
17: return d[m][n]
```

Algorithm 3 describes the steps to compute the string edit distance between two strings. Input to the algorithm is strings A and B, and output is the string edit distance between A and B. Lines 1–2 of the algorithm compute the lengths of the strings A and B. The *for* loop in lines 3–5 denotes the deletion of all characters in A to match the empty string B. The *for* loop in lines 6–8 represents the insertion of all characters in A to match B. When the characters in two strings are same then *if* loop in lines 11–12 shows no change needed for matching characters. Otherwise, the *else* loop in line 13 computes the edit distance using the recurrence relations.

$$d[i][j] = \min(d[i-1][j]+1, d[i][j-1]+1, d[i-1][j-1]+1)$$

In the above *dynamic programming*-based recurrence relation of $d[i][j]$ the first component $d[i][j-1]+1$ denotes insertion operation (insert character B into A); the second component $d[i-1][j]+1$ reflects deletion operations (delete character from A); and the third component $d[i-1][j-1]+1$) represents substitution operations (substitute character in A to match B). Finally, Algorithm 3 returns the minimum number of edit operations required to transform A into B in line 17.

The time complexity of the String-Edit-Distance algorithm is $O(mn)$. Where m is the length of string A and n is the length of string B. The above complexity is due to *for* loops of lines 9 and 10, which execute m and n times, respectively. Space complexity is also $O(mn)$ for creating and storing the values in the dynamic programming table. Space complexity can be reduced to $O(\min(m, n))$ with space optimization.

Damerau-Levenshtein Distance

Damerau-Levenshtein distance is an extension of Levenshtein distance that includes transposition of adjacent characters in addition to insertion, deletion, and substitution of characters [49]. For two strings "abcd" and "acbd", swapping 'b' and 'c' converts "abcd" to "acbd". Therefore, the Damerau-Levenshtein distance between "abcd" and "acbd" is 1.

Definition 2.3 (Damerau-Levenshtein distance) *Damerau-Levenshtein distance* between two strings a and b is defined as:

$$d_{a,b}(i,j) = min \begin{cases} 0, & \text{if } i=j=0 \\ d_{a,b}(i-1,j)+1, & \text{if } i>0 \\ d_{a,b}(i,j-1)+1, & \text{if } j>0 \\ d_{a,b}(i-1,j-1)+1, & \text{if } i,j>0 \text{and } a_i \neq b_j \\ d_{a,b}(i-2,j-2)+1, & \text{if } i,j>1 \text{and } a_{i-1}=b_j, a_i=b_{j-1} \end{cases}$$

although Damerau-Levenshtein distance was initially applied for checking spelling errors, now it is also being used in various domains, including optical character recognition and in bioinformatics, to measure the distance between protein sequences.

Jaro Distance

Jaro distance measures similarity rather than difference, ranging from 0 (completely different) to 1 (identical strings). It considers the number of matching characters along with the number of transpositions of characters that are out of order.

$$J = \frac{1}{3}\left(\frac{m}{|s_1|} + \frac{m}{|s_2|} + \frac{(m-t)}{m}\right)$$

Where s_1 and s_2 are the lengths of the two strings, m is the number of matching characters, and t is the number of transpositions. An enhancement of Jaro distance is *Jaro-Winkler distance*, which gives more weight to matching prefixes. The idea is that typographical errors tend to occur near the end of words rather than at the beginning. Its applications are in search engines and auto-suggestions.

Longest Common Subsequence Distance

Longest common subsequence (LCS) distance measures how many characters appear in both strings without reordering. Unlike edit distance, it only allows deletion operations. LCS distance is equal to the difference of the length of the longer string and LCS length. Let x= *"abcbdab"* and y= *"bdcaba"* be two sequence then LCS of x and y is *"bcba"* having length 4. Then LCS distance = Length of the string x - LCS length =7-4=3.

Cosine Similarity

Cosine similarity measures the cosine of the angle between two strings represented as vectors in a high-dimensional space. This measures is often used for comparing documents or long strings. Cosine similarity is also applied in document similarity for plagiarism detection and natural language processing.

$$\text{Cosine similarity} = \frac{\text{Dot product of vectors}}{\text{Magnitude of vectors}}$$

Table 2.3 provides the comparison of various distance measures for strings. It lists the allowed operations of each distance measures, along with whether the measure is suitable for fixed-length strings. It also lists the computational complexity of each distance measures.

2.4.2 Tree Matching

Tree matching involves comparing two trees to measure their similarity or to find an optimal transformation between them. It is a fundamental problem

TABLE 2.3: Comparison of Distance Measures

Distance Measures	Allowed Operations	Fixed-Length	Complexity
Edit Distance	Insert, Delete, Substitute	No	$O(mn)$
Hamming Distance	Substitutions Only	Yes	$O(n)$
Damerau-Levenshtein	Insert, Delete, Substitute, Transpose	No	$O(mn)$
Jaro-Winkler Distance	Matches, Transpositions, Prefix	No	$O(n)$
LCS	Deletions Only	No	$O(mn)$
Cosine Similarity	Vector-Based Comparison	No	$O(n)$

in SPR, where objects are represented as hierarchical structures (trees) rather than as feature vectors. This approach is particularly useful in applications where data is naturally represented as hierarchical structures. Tree matching techniques can be broadly categorized into exact and inexact tree matching based on the type of constraints and transformations allowed.

Exact Tree Matching

Two trees are considered identical if there exists a one-to-one mapping between their nodes while preserving the hierarchical structure. Exact tree-matching algorithms determine whether two trees are structurally identical while preserving node relationships. Two trees, T_1 and T_2, are ***isomorphic*** if there exists a one-to-one correspondence between the nodes of tree T_1 and tree T_2 while preserving their parent-child relationship. To check two trees for isomorphism, traverse both trees using depth-first search or breadth-first search, and check if both trees have the same number of nodes and identical structural relationships. ***Tree isomorphism algorithm*** works by recursively comparing the structure of both trees while considering node relationships. Algorithm first checks basic properties of the tree; if both trees are empty, they are isomorphic; if the number of nodes differs, they are not isomorphic; if the heights of the trees differ, they are not isomorphic. Compare the degree of corresponding nodes. Recursively compare corresponding subtrees for structural similarity. If all nodes and their respective children match in both trees, they are isomorphic. The complexity of this method is $O(n)$ for trees with n nodes using canonical forms. The subtree isomorphism problem checks if one tree is embedded as a structure within another tree. The algorithm for subtree isomorphism performs recursive traversal of the larger tree. Then it compares each subtree to the given smaller tree using tree isomorphism. The time complexity of this approach is $O(mn)$ for trees with m and n nodes.

Inexact Tree Matching

Inexact or approximate tree matching allows for differences between trees by defining a cost function for transformations such as node insertion, deletion, and substitution.

Definition 2.4 (Tree Edit Distance) *The Tree Edit Distance from T_1 to T_2 is defined by*

$$Tree\ Edit\ Distance(T_1, T_2) = min_{(e_1,\ldots,e_k)\in\varphi(T_1,T_2)} \sum_{i=1}^{k} c(e_i)$$

here $c(e_i)$ represents the costs of corresponding edit operations of e_i and $\varphi(T_1, T_2)$ denotes the sequence of edit path converting T_1 to T_2.

Algorithm 4 describes the steps to compute the tree edit distance between two trees. The Tree-Edit-Distance algorithm uses dynamic programming to compute and store the intermediate results. During each iteration, the algorithm checks if the result is already stored in the table in line 1, and then the algorithm returns the result in line 2. The algorithm recursively computes tree edit distance in lines 13–15 and returns the minimum cost edit path to convert T_1 into T_2 in line 16.

Algorithm 4 : Tree-Edit-Distance (T_1, T_2)

Input: Two Trees T_1, T_2, where $V_1 = \{u_1, ..., u_n\}$ and $V_2 = \{v_1, ..., v_m\}$
Output: A minimum cost TED between T_1 and T_2
1: **if** (results (T_1, T_2) is known) **then**
2:     **return** results (T_1, T_2)
3: **end if**
4: **if** $(T_1 = \phi \wedge T_2 = \phi)$ **then**
5:     **return** 0
6:     **else if** $(T_1 \neq \phi \wedge T_2 = \phi)$
7:     $v_1 \leftarrow$ rightmost node of T_1
8:     **return** Tree-Edit-Distance$(T_1 - v_1, \phi) + c(v_1 \rightarrow \epsilon)$
9:     **else if** $(T_1 = \phi \wedge T_2 \neq \phi)$
10:     $v_2 \leftarrow$ rightmost node of T_2
11:     **return** Tree-Edit-Distance$(\phi, T_2 - v_2) + c(\epsilon \rightarrow v_2)$
12: **end if**
13: $ted_{del} \leftarrow$ Tree-Edit-Distance$(T_1 - v_1, T_2) + c(v_1 \rightarrow \epsilon)$
14: $ted_{ins} \leftarrow$ Tree-Edit-Distance$(T_1, T_2 - v_2) + c(\epsilon \rightarrow v_2)$
15: $ted_{sub} \leftarrow$ Tree-Edit-Distance$(T_1 - v_1, T_2 - v_2) + c(v_1 \rightarrow v_2)$
16: **return** results $(T_1, T_2) \leftarrow min(ted_{del}, ted_{ins}, ted_{sub})$

Approximate subtree matching is a technique used to determine the best possible alignment between a subtree of one tree and a part of another tree while allowing for small variations such as insertions, deletions, and substitutions. It is similar to subtree isomorphism but allows partial matches with modifications. Instead of exact node matching, this technique compute a similarity score between nodes and use dynamic programming to optimize tree alignment. ***Tree alignment algorithm*** is a method used to compare and align two hierarchical structures (trees) by finding their best correspondence while minimizing a given cost function. It extends edit distance by considering node reordering and alignment constraints. This algorithm assign a cost to node reordering, insertions, and deletions. Then it computes an optimal alignment to minimize the total cost. ***Probabilistic tree matching*** is an approach used in structural pattern recognition where tree structures are compared using probabilistic models rather than strict deterministic rules. This method is useful when trees represent uncertain or noisy data, such as in biological phylogenies, natural language processing, or pattern recognition in

TABLE 2.4: Comparison of Tree Matching Algorithms

Algorithm	**Type**	**Complexity**	**Flexibility**	**Use Case**
Tree Isomorphism	Exact	$O(n)$	Low	Graph theory, indexing
Subtree Isomorphism	Exact	$O(mn)$	Low	plagiarism detection
Tree Edit Distance	Approximate	$O(n^3)$	High	NLP, XML comparison
Approx. Subtree Matching	Approximate	$O(mn)$	High	Handwritten recognition
Tree Alignment	Approximate	$O(n^3)$	High	Scene graph matching

images. Unlike deterministic tree matching, which requires exact correspondence, probabilistic methods assign a likelihood score based on node similarity, edge similarity or structural similarity. It computes node similarity using statistical measures like Gaussian distributions or similarity kernels. Edge similarity is computed using transition probabilities for parent-child relationships. Finally it computes structural similarity based on sub-tree alignments and probabilistic models.

Table 2.4 describes comparison of important tree matching algorithms. It lists exact as well as approximate tree matching techniques along with their computational complexity. We can observe in the table that exact matching techniques are less flexible on the other hand approximate matching techniques are more flexible.

2.4.3 Graph Matching

Graph matching is one of the most fundamental technique in SPR, used to determine the similarity or correspondence between two graphs. It involves finding a mapping between the nodes and edges of one graph to those of another while preserving the underlying structural relationships. Graph matching is the process of finding similarity between two graphs. It is classified in to exact and inexact graph matching depending on the matching constraints. For exact graph matching between two graphs there must exists a one-to-one correspondence between nodes and edges of one graph to another one. One of the notable limitations of exact graph matching lies in its stringent criteria for establishing matching. This inflexibility causes exact matching less suitable for real-world applications. Exact graph matching can only determine whether two graphs are identical or not without providing any insight into the degree of their similarity. Inexact graph matching also known as error-tolerant graph matching extends the capabilities of exact matching by quantifying the degree of similarity between graphs that are not perfectly identical.

Definition 2.5 (Graph Isomorphism) *Two graphs* $G_1 = (V_1, E_1)$ *and* $G_2 = (V_2, E_2)$ *are said to be isomorphic, if there exists a bijection* $f : V_1 \to V_2$

such that the edges are preserved:

$$(u, v) \in E_1 \Leftrightarrow (f(u), f(v)) \in E_2.$$

The mapping f is called graph isomorphism and the problem of finding this mapping $f : V_1 \rightarrow V_2$ is called as graph isomorphism problem.

By extending the concept of string edit distance to trees, the tree edit distance was formalized. Similarly, by extending the concept of tree edit distance to graphs, the graph edit distance is conceptualized. In other words graph edit distance is generalization of tree edit distance which in turn is generalization of string edit distance.

Definition 2.6 (Graph Edit Distance) *The Graph Edit Distance (GED) between two graphs G_1 and G_2 is defined as*

$$GED(G_1, G_2) = min_{(e_1,\ldots,e_k)\in\varphi(G_1,G_2)} \sum_{i=1}^{k} c(e_i)$$

Where $\varphi(G_1, G_2)$ is the set of edit distance paths converting G_1 to G_2, and $c(e_i)$ is the cost function associated with edit operation e_i.

Graph edit distance measures the minimum number of edit operations required to transform one graph into another. Allowed edit operations are insertion, deletion and substitution of node and edges. A sequence of edit operations that transform a graph G_1 into G_2 is called an edit path between G_1 and G_2. A common set of edit operations include insertions, deletions, and substitutions of nodes and edges. The notation $\epsilon \rightarrow u$ denotes the insertion of a vertex u, $u \rightarrow \epsilon$ indicates deletion of vertex u, and $u \rightarrow v$ represents substitution of a vertex u by vertex v. Similarly, $\epsilon \rightarrow e$ signify the insertion of an edge e, $e \rightarrow \epsilon$ represents deletion of the edge e, and $e \rightarrow f$ denotes substitution of an edge e by edge f.

Algorithm 5 describes the steps to compute graph edit distance between graphs G_1 and G_2. Input to the Graph-Edit-Distance algorithm is graph G_1 and G_2 along with cost function $cost_{fun}$. Output of the algorithm is graph edit distance between G_1 and G_2. Lines 1–2 of algorithm initializes the values n_1 and n_2 by the number of nodes in graphs G_1 and G_2 respectively. Lines 4–9 represent the cost of transforming an empty graph. Lines 10–21 of the algorithm compute the minimal cost for transforming G_1 into G_2. Computation of minimum cost considering node is performed in line 15 and line 21 updates with minimum cost considering edges. Finally, line 26 returns the minimum cost to transform G_1 into G_2.

Time complexity of the Graph-Edit-Distance algorithm is exponential due to the very large number of possible edit path sequences. Due to exponential complexity of graph edit distance computation several heuristic and approximation algorithms have been proposed.

Algorithm 5 : Graph-Edit-Distance $(G_1, G_2, cost_{fun})$

Input: Two Graphs $G_1 = (V_1, E_1)$ and $G_2 = (V_2, E_2)$
Output: Graph edit distance between G_1 and G_2

```
n1 ← |V1|
n2 ← |V2|
D[0][0] ← 0
for (i ← 1 to n1) do
    D[i][0] ← D[i − 1][0] + cost_fun.del_node(V1[i])
end for
for (j ← 1 to n2) do
    D[0][j] ← D[0][j − 1] + cost_fun.ins_node(V2[j])
end for
for (i ← 1 to n1) do
    for (j ← 1 to n2) do
        cost_del ← D[i − 1][j] + cost_fun.del_node(V1[i])
        cost_ins ← D[i][j − 1] + cost_fun.ins_node(V2[j])
        cost_sub ← D[i − 1][j − 1] + cost_fun.sub_node(V1[i], V2[j])
        D[i][j] ← min(cost_del, cost_ins, cost_sub)
        for each edge (V_ik, V_il) ∈ E1 do
            for each edge (V_jm, V_jn) ∈ E2 do
                e_del ← D[i − 1][j] + cost_fun.del_edge(V_ik, V_il)
                e_ins ← D[i][j − 1] + cost_fun.ins_edge(V_jm, V_jn)
                e_sub ← D[i − 1][j − 1] + cost_fun.sub_edge((V_ik, V_il), (V_jm, V_jn))
                D[i][j] ← min(D[i][j], cost_del, cost_ins, cost_sub)
            end for
        end for
    end for
end for
return D[n1][n2]
```

2.5 Learning in Structural Pattern Recognition

SPR focuses on recognizing complex patterns based on their structural relationships rather than only their feature-based properties. Learning in SPR involves extracting, generalizing, and adapting structural representations from training data to improve classification and recognition accuracy. Machine learning techniques are increasingly being applied to structural pattern recognition, allowing systems to learn patterns from data rather than relying on hand-crafted rules.

Supervised Learning

Supervised learning in SPR refers to the process of learning a mapping from structured input data (e.g., graphs, trees, strings) to output labels using a set of labeled training examples. The goal is to generalize from known patterns and apply the learned knowledge to new, unseen structured data. In supervised learning for SPR, the goal is to learn a mapping from structured inputs to labels or outputs. The system is trained using labeled structured data such as graphs, trees and strings.

Key steps of supervised learning in SPR include training data, feature representation, learning algorithms, and predictions or classification. Data is trained using a set of structured objects with known class labels. For feature extractions, structural data must be represented in a way that enables learning. Common representations for feature extractions are edit distance, graph kernel, and embedding. Learning algorithms allow models to learn from labeled structural data. Common learning algorithms are graph neural networks, support vector machines, and decision trees. During the classification stage, the model predicts the class of a given structured pattern based on the learned patterns.

Graph neural network is a powerful deep learning approach that extends traditional neural networks to handle graph-structured data, enabling end-to-end learning from graph-based representations. It can also be used to learn embedding from graph structures. Edit distance-based classifiers can use graph edit distance for classification. Kernel methods such as graph kernels can apply various kernel functions to compare structured data. Supervised learning enables powerful pattern recognition in structured data, but its success depends on effective feature representation, computational efficiency, and high-quality labeled datasets.

Unsupervised Learning

Unsupervised learning in structured data refers to learning patterns, clusters, or representations from structured objects without labeled data. The goal is to discover inherent structures, similarities, and relationships within the data. The important techniques of unsupervised learning in structured data are clustering, dimensionality reduction, frequent pattern mining, and feature learning. ***Clustering*** groups similar structured objects based on shared properties. Common methods of clustering for structural data include graph clustering, spectral clustering, and hierarchical clustering. ***Graph clustering*** is the task of partitioning the nodes of a graph into groups like clusters or communities based on their structural connectivity or similarity. The goal is to identify meaningful substructures within a graph where nodes in the same cluster are more closely connected to each other than to nodes in other clusters. Partition-based class clustering divides the graph into k non-overlapping clusters. For example, spectral clustering uses eigenvalues of the graph Laplacian. Hierarchical clustering builds a tree of clusters, either by merging smaller clusters or splitting larger ones. Density-based clustering detects dense subgraph clusters separated by sparse regions. ***Dimensionality reduction*** reduces the complexity of structured data while preserving meaningful information. Common methods of dimensionality reduction are kernel principal component analysis and graph embedding. Kernel principal component analysis extends principal component analysis to structured data. Graph embedding converts graphs into low-dimensional vectors. ***Frequent pattern mining*** identifies the recurring substructures in structured data, such as finding common subgraphs in chemical compounds. Feature learning enables to learn meaningful structural representation from data. Unsupervised learning in structured data enables pattern discovery without predefined labels, making it a powerful tool for exploratory data analysis and representation learning.

Semi-Supervised Learning

Traditional machine learning approaches often rely on either fully supervised or unsupervised methods. However, in many real-world applications, labeled data is scarce and expensive to obtain, while large amounts of unlabeled data are readily available. Semi-supervised learning provides a powerful paradigm for leveraging both labeled and unlabeled data to improve classification and recognition performance in structural domains. Semi-supervised learning operates between supervised and unsupervised learning by utilizing a small set of labeled examples along with a larger set of unlabeled examples. The key assumption in semi-supervised learning is that the underlying data distribution provides useful information about class structure, which can be exploited to improve learning.

Several semi-supervised learning methods have been adapted for structural pattern recognition, leveraging graph-based models and relational learning techniques. Graph-based semi-supervised learning is particularly relevant for structural pattern recognition, as graphs naturally capture relationships between data points. Label propagation spreads label information from labeled to unlabeled nodes in a graph based on similarity measures. Graph neural networks learn node embeddings using labeled and unlabeled data, incorporating structural dependencies. Graph-based regularization encourages similar nodes in a graph to have similar predictions by adding a smoothness constraint. Self-training approaches iteratively assigns pseudo-labels to unlabeled data, retraining the model with a mix of labeled and pseudo-labeled samples. This is useful for structured data when reliable confidence measures can be defined. Graph-based generative models enable to model complex structural dependencies to create realistic representations for semi-supervised training. Generative models such as variational autoencoders learn latent representations of structured data and generate new samples.

Semi-supervised learning provides a promising approach for structural pattern recognition, leveraging both labeled and unlabeled data to enhance learning performance. By integrating graph-based techniques, self-training, and generative models, semi-supervised learning has demonstrated effectiveness in a range of structural applications.

Reinforcement Learning

Reinforcement Learning is a machine learning paradigm in which an agent learns to make sequential decisions by interacting with an environment to maximize cumulative rewards. While reinforcement learning has been extensively applied to tasks like robotics, game playing, and control systems, its application to structured data is an emerging area of research. Structured data presents unique challenges due to its inherent relational and non-Euclidean properties. Adapting reinforcement learning to structured data requires specialized techniques that can capture dependencies and hierarchical relationships effectively.

Sequence-based reinforcement learning approach is applied to sequential decision problems in recurrent neural network for capturing long-term dependencies in structured sequences. Attention-based models like transformer-based reinforcement learning (e.g., Decision Transformer) improve learning from structured sequential data. Tree structures are common in hierarchical decision processes and structured data representations. Reinforcement learning leverages tree-based models in Monte Carlo tree search, which is a planning algorithm that balances exploration and exploitation by simulating future action sequences. Hierarchical reinforcement learning is a technique where high-level policies make abstract decisions, and low-level policies refine them. Graphs naturally represent structured data in applications like knowledge graphs, molecular structures, and social networks. Reinforcement learning can be applied to graphs through graph neural network and graph-based exploration. Graph neural network such as graph convolutional network and graph attention network are used to model graph-structured environments. Graph neural network encode graph relationships and pass messages between nodes to improve decision-making.

Reinforcement Learning provides a promising framework for handling structured data in graphs, trees, and sequences. Advances in graph-based reinforcement learning, hierarchical learning, and deep reinforcement learning models have led to applications in drug discovery, NLP, recommender systems, and more. Continued research in scalable RL algorithms, better state representations, and efficient exploration strategies will drive further improvements in structured data learning..

2.6 Applications of Structural Pattern Recognition

SPR is used in a variety of fields where structured data representations are necessary to capture complex relationships and patterns.

2.6.1 Image and Object Recognition

In image analysis, SPR is used to recognize objects based on their structural relationships, rather than relying solely on pixel-based features.

- **Shape Recognition:** Graph-based representations of shapes (e.g., skeletons or contours) are used to recognize objects in images.
- **Scene Understanding:** SPR techniques can be applied to understand the relationships between objects in a scene, such as their spatial arrangement.

2.6.2 Natural Language Processing

In natural language processing (NLP), SPR is used to model the syntactic and semantic structure of language.

- **Parsing:** Tree-based representations are used to parse the syntactic structure of sentences, which can then be used for translation, summarization, or information extraction.
- **Text Mining:** Graph-based representations of text documents (e.g., word co-occurrence graphs) are used for tasks such as document clustering or keyword extraction.

2.6.3 Bioinformatics

SPR plays a critical role in bioinformatics, where biological data often exhibits complex structural relationships.

- **Protein Structure Analysis:** Proteins are represented as graphs of atoms connected by chemical bonds, and graph matching techniques are used to compare protein structures.
- **Genomic Sequence Matching:** String matching and graph-based methods are used to compare DNA sequences and study evolutionary relationships.

2.6.4 Social Network Analysis

Social networks are naturally represented as graphs, where nodes represent individuals and edges represent relationships. SPR techniques are used to identify patterns in these networks.

- **Community Detection:** Graph clustering methods are used to identify groups of closely connected individuals within a social network.

- **Role Identification:** Graph-based pattern recognition can identify important actors in social networks, such as influencers or bridge nodes between communities.

The applications of SPR are broad and diverse, spanning multiple domains. Here are a few notable examples:

- **Bioinformatics and Molecular Biology:** In the study of biological systems, understanding the structural patterns within DNA, RNA, and protein sequences is prominent. SPR techniques are employed to identify motifs, predict protein folding, and analyze genetic relationships.

- **Natural Language Processing:** Languages possess complex structural hierarchies. SPR plays a crucial role in tasks like parsing sentences, recognizing parts of speech, and syntactic analysis, enabling machines to understand and generate human language.

- **Computer Vision:** While traditional pattern recognition in computer vision focuses on pixel-level features, SPR extends this to higher-level structures. It enables the recognition of objects in images, understanding scene layouts, and even interpreting 3D shapes and their relationships.

- **Handwriting and Signature Verification:** In forensics and document analysis, SPR is used to identify the structural nuances in handwriting and signatures. It aids in authentication and verification processes.

- **Graph and Network Analysis:** In domains like social network analysis, transportation networks, and biological networks, data often takes the form of graphs. SPR methods are employed to discover patterns in these complex, interconnected systems.

2.7 Challenges in Structural Pattern Recognition

Despite its power, structural pattern recognition comes with several challenges that researchers and practitioners must address.

2.7.1 Computational Complexity

Many SPR problems, such as graph matching and parsing, are computationally expensive, particularly for large datasets.

- **Scalability:** Developing efficient algorithms that scale to large, real-world datasets is a critical challenge.

- **Approximate Methods:** Approximate algorithms, such as greedy matching or heuristic search, are often employed to reduce computational costs while maintaining acceptable accuracy.

2.7.2 Handling Noise and Incomplete Data

Many SPR problems, such as graph matching and parsing, are computationally expensive, particularly for large datasets.

- **Robustness:** SPR algorithms must be robust to variations in the data, such as missing nodes or edges in a graph.
- **Probabilistic Models:** Probabilistic approaches can help handle uncertainty by assigning likelihoods to different patterns or matches.

2.7.3 Interpretability

Structured models often produce complex outputs that are difficult to interpret, especially when machine learning techniques are involved.

- **Explainability:** Ensuring that the patterns recognized by an SPR system are interpretable and understandable is crucial, especially in domains like healthcare or security.

Graphs are a fundamental structure in SPR. Graph matching algorithms aim to compare and align graphs to discover similarities. It is crucial in applications like image recognition, where objects can be represented as graphs. Dynamic programming is a powerful technique used in SPR to efficiently solve problems by breaking them down into overlapping subproblems. It's particularly effective for tasks involving sequences or strings. Hierarchical clustering classifies data points based on their similarity or dissimilarity. This is instrumental in uncovering hierarchical structures within datasets. With the advent of deep learning, neural networks have been adapted for SPR tasks. Convolutional neural networks (CNNs) are used for tasks like image recognition, while recurrent neural networks (RNNs) are employed for sequential data. Graph neural networks (GNNs) are a specialized form of neural network designed to handle graph-structured data. They have shown remarkable success in tasks involving graph representation learning and node classification.

While SPR has made significant strides, it still faces challenges in dealing with noisy or incomplete data, as well as in scaling to handle extremely large datasets. Additionally, interpreting and visualizing the learned structures remains an active area of research. Looking ahead, the integration of SPR with other fields like reinforcement learning, natural language processing, and computer vision holds great promise. The development of more interpretable and explainable models is also a crucial avenue for further research.

Structural Pattern Recognition stands as a cornerstone in the field of artificial intelligence, providing a powerful framework for understanding and interpreting complex, structured data. Its applications are far-reaching, from deciphering genetic sequences to parsing natural language. With the continued advancement of techniques and algorithms, SPR is expected to play an increasingly pivotal role in unraveling the intricate patterns that permeate our world.

Chapter 3

Graph Matching Algorithms: A Survey

3.1 Introduction

Graph matching is the process of computing the similarity between two graphs. This process is essential in various domains, including computer science and pattern recognition. The categorization of graph matching into exact and inexact graph matching shows distinct theoretical and practical implications. The exact graph matching, while having substantial theoretical significance, poses complex challenges in terms of computational complexity. In contrast, inexact graph matching holds significant practical relevance in real-world applications, addressing scenarios where input data is subject to noise and distortion.

For instance, the graph isomorphism problem, an example of exact graph matching, has deep implications in complexity theory. The determination of its complexity class, whether it resides in class P or is NP-complete, remains unresolved. On the other hand, the subgraph isomorphism problem, an instance of inexact graph matching, is known to be NP-complete. Its NP-completeness underlines the absence of an efficient polynomial-time algorithm for inexact graph matching.

This chapter presents a comprehensive review of various graph matching techniques, encompassing exact, approximate, and error-tolerant approaches. A seminal work by Conte *et al.* [37] provides an exhaustive survey of exact and inexact graph matching techniques used in pattern recognition. Building on this, Foggia *et al.* [71] present a more recent survey, delving into graph matching and learning techniques pivotal to pattern recognition. Additionally, Bunke's paper [22] introduces a formal framework and algorithmic foundation for graph matching, contributing significantly to this field. Gao *et al.*'s survey [76] offers a comprehensive overview of algorithms addressing the graph edit distance problem. Further insights into the graph matching problem, with a focus on methodological and algorithmic results, are provided by Livi and Rizzi in their review [123]. Dwivedi [56] presents algorithms on exact, approximate, and error-tolerant graph matching.

DOI: 10.1201/9781003532248-3

This chapter lays the foundation by presenting fundamental definitions and concepts essential to both exact and error-tolerant graph matching. It establishes a common terminology and knowledge necessary for the subsequent exploration of various matching techniques. Here, we explore an in-depth review of exact graph matching techniques. It provides a comprehensive overview of approaches and methodologies needed in scenarios where an absolute correspondence between vertices and edges is sought. Then, we shift our focus towards error-tolerant graph matching methods. These techniques play a crucial role in scenarios where flexibility and tolerance to noise and distortion in input data are imperative.

This chapter is organized as follows. Section 3.2 presents basic definitions and concepts used in exact and error-tolerant graph matching. Section 3.3 describes a survey of exact graph matching techniques. Section 3.4 presents a review of error-tolerant graph matching methods. Finally, section 3.5 describes the summary.

3.2 Basic Concepts and Definitions

In this section, we review the basic concepts and definitions used in exact and inexact graph matching [1, 138, 147, 156].

Definition 3.1 (Graph) *A graph G is defined as $G = (V, E, \mu, \nu)$, where*

- *V is the set of vertices or nodes,*
- *E is the set of edges or links,*
- *$\mu : V \rightarrow L_V$ assigns a node label set $l_v \in L_V$ to each vertex $v \in V$,*
- *$\nu : E \rightarrow L_E$ assigns an edge label set $l_e \in L_E$ to each edge in E,*
- *L_V and L_E are node label set and edge label set respectively.*

A graph with labels on its nodes or edges is termed a labeled graph. If $L_V = L_E = \emptyset$ then G is called the unlabeled graph. Figure 3.1 shows examples of labeled and unlabeled graphs. $|G|$ denotes the number of nodes in a graph G.

Definition 3.2 (Subgraph) *Given two graphs $G_1 = (V_1, E_1, \mu_1, \nu_1)$ and $G_2 = (V_2, E_2, \mu_2, \nu_2)$, G_1 is a subgraph of G_2, if*

- *$V_1 \subseteq V_2$,*
- *$E_1 \subseteq E_2$,*
- *$\mu_1(u) = \mu_2(u)$ for every node $u \in G_1$,*

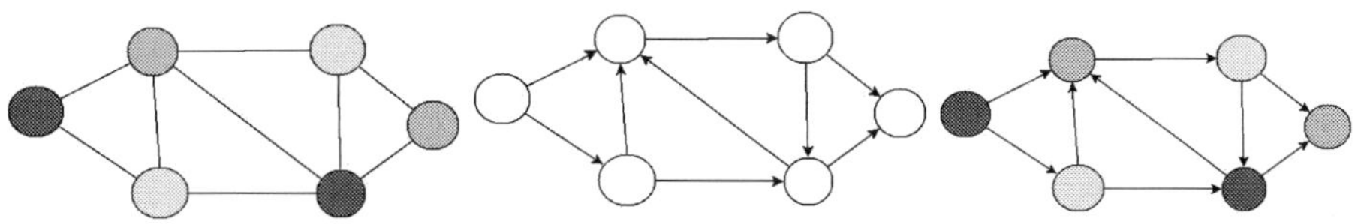

(a) Undirected labeled (b) Directed unlabeled (c) Directed labeled

FIGURE 3.1: Graph Examples

- *$\nu_1(e) = \nu_2(e)$ for every edge $e \in G_1$.*

Definition 3.3 (Graph Isomorphism) *Let $G_1 = (V_1, E_1, \mu_1, \nu_1)$ and $G_2 = (V_2, E_2, \mu_2, \nu_2)$ be two graphs. A graph isomorphism between G_1 and G_2 is a function $f : V_1 \to V_2$ such that*

- $\mu_1(u) = \mu_2(f(u)), \forall u \in V_1$,
- $\forall e_1 = (u, v) \in E_1, \exists e_2 = (f(u), f(v)) \in E_2$ *such that* $\nu_1(e_1) = \nu_2(e_2)$,
- $\forall e_2 = (u, v) \in E_2, \exists e_1 = (f^{-1}(u), f^{-1}(v)) \in E_1$ *such that* $\nu_1(e_1) = \nu_2(e_2)$,

In simple terms, graph isomorphism between two graphs preserves labels and connecting edges via a bijection between their vertices.

Definition 3.4 (Subgraph Isomorphism) *For two graphs G_1 and G_2, a mapping $f : V_1 \to V_2$ from G_1 to G_2 is a subgraph isomorphism if there exists a graph isomorphism between G_1 and a subgraph of G_2.*

Definition 3.5 (Maximum Common Subgraph) *Let G_1 and G_2 be two graphs. A graph G is a common subgraph if it is subgraph isomorphic to both G_1 and G_2. G is the Maximum Common Subgraph (mcs) if there is no larger graph that is a common subgraph to both G_1 and G_2.*

Definition 3.6 (Minimum Common Supergraph) *Let G_1 and G_2 be two graphs. A graph G is a common supergraph if both G_1 and G_2 are subgraph isomorphic to G. G is the Minimum Common Supergraph (MCS) if there is no smaller graph that is a common supergraph to both G_1 and G_2.*

A sequence of edit operations that transform a graph G_1 into G_2 is called an edit path between G_1 and G_2. A common set of edit operations include insertions, deletions, and substitutions of nodes and edges. The notation $\epsilon \to u$ denotes the insertion of a vertex u, $u \to \epsilon$ indicates deletion of vertex u, and $u \to v$ represents substitution of a vertex u by vertex v. Similarly, $\epsilon \to e$ signify the insertion of an edge e, $e \to \epsilon$ represents deletion of the edge e, and $e \to f$ denotes substitution of an edge e by edge f.

Definition 3.7 (Graph Edit Distance) *The Graph Edit Distance (GED) between two graphs G_1 and G_2 is defined as*

$$GED(G_1, G_2) = min_{(e_1,\ldots,e_k)\in\varphi(G_1,G_2)} \sum_{i=1}^{k} c(e_i)$$

Where $\varphi(G_1, G_2)$ is the set of edit distance paths converting G_1 to G_2, and $c(e_i)$ is the cost function associated with edit operation e_i.

GED measures the minimum number of edit operations required to transform one graph into another. It extends tree edit distance and string edit distance by allowing insertions, deletions, and substitutions of nodes and edges.

3.3 Exact Graph Matching

Exact graph matching involves establishing a one-to-one correspondence between the vertices and edges of two graphs, a problem commonly referred to as the graph isomorphism problem. It's crucial to differentiate between "graph matching" and "matching" in graph theory. While graph matching involves comparing two distinct graphs, matching in a graph refers to a set of edges where no two edges share a common vertex. Essentially, matching is defined within a single graph, whereas graph matching requires a comparison between two separate graphs.

In graph theory, an isomorphism signifies a bijective mapping between the vertex sets of two graphs. If such a mapping exists, we consider the graphs to be isomorphic. The graph isomorphism problem is a foundational problem of computational complexity theory. Although it falls within the domain of NP problems, it remains uncertain whether it can be solved in polynomial time or if it is NP-complete. Remarkably, this problem, alongside the renowned integer factorization problem, constitutes one of the few problems outlined by Garey and Johnson in their influential work [77], both of which still elude their complexity resolution. Current conjectures position the graph isomorphism problem in the domain of NP-intermediate problems, which comprise those within NP that aren't proven to be in P or NP-complete. Ladner's [115] demonstration asserted that if $P \neq NP$, NP-intermediate problems must exist. In simpler terms, should P and NP prove unequal, there will undoubtedly be problems categorized as NP-intermediate. Conversely, if P=NP is proven, then NP-intermediate would remain devoid of such problems.

At present, the fastest acknowledged theoretical algorithm for the graph isomorphism problem is attributed to Babai and Luks [7], having a runtime of $2^{O(\sqrt{n \log n})}$ for graphs with n vertices. Babai introduced a quasipolynomial time algorithm with a runtime of $2^{O((\log n)^c)}$ for some fixed $c > 0$, designed for the graph isomorphism problem. Although the overall complexity of the general graph isomorphism problem remains unresolved, polynomial-time algorithms have been devised for specific graph classes. For instance, Luks [126], outlined a polynomial algorithm for bounded valence graphs, and Zemlyachenko *et al.* [199] provided an insightful compilation of work associated with the graph isomorphism problem. Kobler *et al.* [109], summarized the findings of the graph isomorphism problem and its structural complexity.

The subgraph isomorphism problem stands as a generalization of the graph isomorphism problem. Here, the objective is to ascertain whether a graph has a subgraph that mirrors another graph. Essentially, the graph isomorphism problem can be seen as a special instance of the subgraph isomorphism problem. Given that the subgraph isomorphism problem extends from the Hamiltonian cycle problem, it has been classified as NP-complete [39]. Due to the lack of a polynomial time solution, various approximation and suboptimal algorithms have been introduced to tackle the subgraph isomorphism problem. In this section, we will review some of the more practical algorithms designed for exact graph matching.

3.3.1 Tree Search-Based Techniques

Tree search-based methods have been foundational in the field of graph matching. In a seminal paper, Hart *et al.* [90] introduced the A^* search algorithm, a technique designed to uncover the optimal path with the minimum cost between two nodes within a graph. The algorithm hinges on an evaluation function, denoted as $f(n)$, which aggregates two key components: $g(n)$ representing the actual cost from the source node s to n, and $h(n)$ signifying the actual cost from n to the target node t. Through a suitable selection of $f(n)$, the A^* search algorithm unfailingly identifies an optimal path from the source to the target node. For an in-depth exploration of this algorithm and its variants, readers may refer to Nilsson's comprehensive book [142].

Ullmann [186], introduced an algorithm for subgraph isomorphism that extends to both graph isomorphism as well as monomorphism. He devised a refinement procedure to systematically eliminate successor nodes inconsistent with prior matches. In a significant update, Ullmann [187] advanced the field by employing binary constraint satisfaction for subgraph isomorphism. This approach leverages a novel bit-vector algorithm, placing greater emphasis on search and minimizing domain editing. In a noteworthy paper, Ghahraman *et al.* [80] proposed a graph monomorphism algorithm utilizing netgraph derived from the Cartesian product of two graphs' vertices. This innovative approach effectively prunes the search space.

Cordella *et al.* [40] introduced the VF algorithm, employing a state-space representation for graph matching. States here embody partial solutions, with transitions signifying the inclusion of a pair of matched nodes. Stringent criteria are employed to trim partial matching states that don't contribute to the desired graph matching outcome. Subsequent experiments, as detailed in Cordella *et al.*'s work [43], demonstrated the superior performance of the VF algorithm compared to Ullmann's approach. Building on this foundation, Cordella *et al.* [44] introduced the VF2 algorithm. This modified version proved particularly suitable for large graphs, thereby drastically reducing memory requirements in relation to the number of vertices. Larrosa and Valiente [117] provided a novel formulation of the graph isomorphism problem within the framework of constraint satisfaction. Their proposed algorithm exploits the problem's structure to enhance lookahead procedures.

McGregor [129] presented a backtrack search algorithm tailored for the maximal common subgraph problem. Koch's paper [110] outlined a method for enumerating all connected maximal common subgraphs between two graphs, effectively transforming the problem into a clique problem. Bunke *et al.* [25] delved into algorithms for maximum common subgraph detection, with a thorough performance evaluation.

Konc and Janezic's [111] paper introduced a refined approximate coloring algorithm, instrumental in computing the maximum clique problem, thereby helping in the identification of the maximum common subgraph. Their improvement centered on calculating precise upper bounds, particularly at the root of the branch-and-bound algorithm's recursion tree. Zampelli *et al.* [198] proposed a constraint programming approach to the subgraph isomorphism problem. This method incorporated a global constraint and an associated filtering algorithm. Furthermore, Solnon's paper [178] outlined a novel filtering algorithm grounded in local all-different constraints. This technique demonstrated superior efficiency in branch pruning compared to existing methods.

3.3.2 Algebraic Graph Theory Based Techniques

McKay's pioneering work [130] introduced a groundbreaking algorithm called nauty, which remains a foundation in the domain of exact graph matching and graph isomorphism. This algorithm draws upon the rich field of group theory, providing an efficient method for computing the generators of a graph's automorphism group. This automorphism group, in turn, plays a pivotal role in determining the canonical labeling of input graphs. This labeling serves as a blueprint for identifying sets of isomorphic graphs, accomplished by scrutinizing the adjacency matrices in their canonical forms. It's worth noting, however, that in the worst-case scenario, the computation of canonical labeling can exhibit exponential complexity. De Santo *et al.* [52] conducted a comprehensive study, highlighting that the nauty program may demand more computational time compared to alternative algorithms like VF2. Their research encompassed an experimental comparison of four precise graph matching algorithms: Ullmann's algorithm, Schmidt and Druffel's approach, the VF2 algorithm, and Nauty. To facilitate this benchmarking endeavor, the authors meticulously constructed a robust database comprising an extensive array of graphs.

Darga *et al.* [50] introduced saucy, a novel symmetry detection tool that leverages a sparse data structure for the implementation of the automorphism group. This technique demonstrated remarkable efficiency, surpassing nauty for various practical graph datasets. Building upon this success, the paper by Darga *et al.* [51] further refined the saucy program. They improved upon a symmetry discovery algorithm that exploits the inherent sparsity present in both input and output, thereby enhancing its performance. McKay and Piperno [131] advanced the field with their improved iteration of the nauty program, complemented by an ingenious technique dubbed Traces. This technique, distinguished by its efficiency, outperforms many existing graph isomorphism tools.

The contributions of these researchers have significantly advanced the landscape of exact graph matching and graph isomorphism, underscoring the pivotal role played by algorithms rooted in group theory and advanced data structures.

3.3.3 Special Type of Graphs

Despite the NP-completeness of the subgraph isomorphism problem and the yet undetermined status of the graph isomorphism problem within the realm of P and NP-complete, polynomial time algorithms have been developed for a range of special classes of graphs.

Aho *et al.* [2] presented an algorithm for tree isomorphism that operates in polynomial time, marking a significant advancement in the field. Hoffmann and O'Donnel [93] extended this progress by introducing several algorithms for tree pattern matching, providing a thorough analysis of time and space complexities in their paper. Bille's work [15] offered an extensive survey encompassing tree edit distance, alignment distance, and inclusion problems, shedding light on critical metrics for evaluating tree-based structures. Demaine *et al.* in their contribution presented an optimal decomposition algorithm for tree edit distance, boasting a worst-case time complexity of $O(n^3)$, where n represents the size of the trees involved. Akutsu *et al.* [4] proposed exact algorithms tailored for computing tree edit distance for unordered trees, further expanding the algorithmic tools for specialized tree structures.

Hopcroft and Wong [96] introduced a linear time algorithm for identifying isomorphisms between planar graphs, thereby substantially enhancing the efficiency of solving planar graph isomorphism. This technique streamlined the solution from $O(n.logn)$ to linear time, revolutionizing the landscape of planar graph analysis. Luks' [126] revealed a polynomial-time algorithm for the graph isomorphism problem in the context of bounded valence graphs. The author has performed a polynomial-time reduction from the graph isomorphism problem of bounded valence to the color automorphism problem of groups having the composition factor of bounded order. The color automorphism problem for a group consists of evaluating generators for the subgroup of the group that stabilizes the color classes, and this problem is solvable in polynomial time.

Dickinson *et al.*'s methodological breakthrough [53] presented an approach for determining graph isomorphism in a class of graphs endowed with unique node labels, all within polynomial time. They further demonstrated that subgraph isomorphism, maximum common subgraph, and graph edit distance computations can be accomplished in quadratic time. Kuramochi and Karypis [114] unveiled an algorithm capable of computing graph isomorphism for geometric graphs in polynomial time, opening new avenues for geometric graph analysis. These algorithmic strides have collectively propelled the field forward, showcasing the power of specialized approaches in handling specific graph classes.

3.3.4 Other Techniques

Messmer and Bunke [133] introduced a novel approach to detecting graph and subgraph isomorphisms, particularly from a sample graph to an extensive dataset of model graphs. Their technique involved a preprocessing step to construct a decision tree for model graphs, facilitating the identification of subgraph isomorphism from the candidate input graph to model graphs within polynomial time. However, a notable drawback of this method was the exponential computational time required for the preprocessing step, as well as the exponential space needed for decision tree construction in relation to the number of nodes in the graphs. Weber *et al.* [193] extended this approach, focusing on reducing storage requirements and indexing time for graphs significantly. Building upon Messmer and Bunke's initial work, they refined the methodology to enhance efficiency. Messmer and Bunke [134] proposed another innovative technique for finding subgraph isomorphism, this time between an input graph and an extensive database of pre-processed graphs. This method streamlined the process by representing frequently occurring subgraphs only once, thereby saving execution time when locating them in the input graph.

Irniger and Bunke [98] introduced decision tree structures tailored for graph database filtering. Their approach involved representing graphs using feature vectors, which were then utilized to construct a decision tree for indexing the graph database. This innovative indexing technique marked a significant advancement in graph database management. Gori *et al.* [84] laid out a comprehensive framework for both exact and approximate graph matching, leveraging a random walk approach based on Markovian graph spectral analysis. They presented a polynomial time algorithm specifically designed for solving the graph isomorphism problem for Markovian spectrally distinguishable graphs. These were defined as a set of graphs that resisted easy reduction to others. Through extensive experiments, the authors conclusively demonstrated the remarkable efficiency of their proposed algorithm in comparison to existing techniques. Furthermore, this concept proved to be highly effective for approximate graph matching scenarios.

Dahm *et al.* [48] introduced a refined approach to subgraph isomorphism detection. Their method involved strategically eliminating nodes based on their topological features. This technique was particularly advantageous for large graphs, streamlining the process by removing nodes from the second graph that were incompatible with nodes from the first graph based on their topological attributes before subjecting them to subgraph isomorphism testing. This advancement represented a significant leap forward in the field of subgraph isomorphism detection.

3.4 Error-Tolerant Graph Matching

One of the notable limitations of exact graph matching lies in its stringent criteria for establishing matching. This inflexibility causes exact matching less suitable for real-world applications. Specifically, exact matching can only determine whether two graphs are identical or not, without providing insight into the degree of their similarity. It means that exact matching fails to explore the range of similarities between isomorphic graphs, which are precisely similar, and non-isomorphic graphs, which are fundamentally dissimilar.

In practical scenarios, input graph data may undergo alterations due to noise introduced during storage and acquisition, or deformation during the process of graph extraction from objects. In such cases, exact matching proves inadequate for differentiating matches or similarities between two graphs. Consequently, error-tolerant graph matching emerges as a practical alternative, enabling general matching between graphs while also accommodating graph and subgraph isomorphism checks. Error-tolerant matching thus extends the capabilities of exact matching by quantifying the degree of similarity between graphs that are not perfectly identical.

Error-tolerant matching, often referred to as inexact matching, excels at finding approximate matches between graphs within a reasonable time frame, particularly in situations where determining the optimal solution is computationally expensive and impractical. This makes error-tolerant matching valuable even in cases where input data distortion is not a concern, as it yields an approximate yet efficient match when identifying the best solution is unfeasible.

Furthermore, error-tolerant matching is alternatively known as error-correcting graph matching, owing to its capacity to rectify distortions, such as deformations, arising from noise or other forms of interference during graph processing. Numerous techniques for error-tolerant graph matching revolve around different types of errors, each associated with its respective cost. The objective is to find a matching that incurs the minimum error. One prevalent class of techniques involves graph edit operations, encompassing operations like node and edge insertions, deletions, and substitutions, each with its own associated cost. The graph edit cost is then defined as the minimum cost of a set of edit operations required to transform one graph into another.

In the subsequent sections of this discourse, we undertake a survey of the pivotal techniques in error-tolerant graph matching as proposed in the literature. This comprehensive examination sheds light on the diverse approaches and methodologies devised to address the challenges posed by inexact matching scenarios.

3.4.1 Tree Search-Based Techniques

Tsai and Fu [184] introduced an innovative approach known as error-correcting isomorphism of attributed relational graphs for matching deformed patterns. This method leverages an ordered search tree-based algorithm, combining structural techniques via pattern representation as a relational graph, and statistical techniques using deformation probabilities to enable error-tolerant matching. Building upon this, they [185] extended the concept to error-correcting subgraph isomorphism, stating it as a state-space tree search problem. The authors also presented an ordered search algorithm for optimal error-tolerant subgraph isomorphism.

Ghahraman *et al.* devised an algorithm [80] targeting the optimal graph monomorphism problem. This was accomplished by defining graph morphisms through subgraphs of a Cartesian product graph and employing a branch-and-bound algorithm for efficient computation. Shapiro and Haralick proposed relational homomorphism [171], a method facilitating structural description comparison through effective tree-search techniques. Sanfeliu and Fu put forth a novel distance measure [160] for non-hierarchical attributed relational graphs. This measure quantifies the minimum number of modifications necessary to transform one graph into another. To ascertain the cost of recognition of nodes, they introduced distinct cost functions for vertices, pivotal in determining similarity between nodes.

Eshera and Fu introduced a novel technique [67] for computing a global distance measure between attributed graphs, specifically for image analysis. Their method involved the decomposition of attributed relational graphs into basic attributed relational graphs, essentially one-level trees. Subsequently, a state-space representation was derived from these basic graphs to facilitate optimal matching.

Cordella *et al.* contributed an efficient algorithm [41] for error-tolerant matching of ARG graphs, employing a well-defined set of syntactic and semantic transformations. This paper contextualized transformations in relation to a prototype, providing applicability when the sample graph aligns with the prototype. The same authors extended their work in [42], employing state space representation to demonstrate partial matching between two graphs. Serratosa *et al.* [170], proposed an efficient algorithm to find a suboptimal similarity measure between Function Described Graph (FDG) and ARG using tree search, where FDG itself is depicted as the ensemble of ARG's.

Berretti *et al.* provided an efficient solution [13] for indexing and matching of ARGs, utilizing a heuristic approach based on A^* search. In a subsequent paper, they further enhanced their approach [14], addressing subgraph error-correcting isomorphism through a look-ahead technique for computing object distances. Gregory and Kittler proposed [86] color image retrieval using graph search techniques, incorporating matching through graph edit operations and optimal search methods.

Sanfeliu *et al.* expanded their work [159] on function-described graphs, detailing a representation for a set of attributed graphs, termed function-described graphs. They also introduced a distance measure for matching attributed graphs with function-described graphs, along with techniques for image segmentation and object recognition. In their paper, Sanfeliu *et al.* [161] investigated object learning and recognition using second-order random graphs, encompassing both marginal and second-order joint probability functions of graph elements. Serratosa *et al.* elaborated [168] on both supervised and unsupervised synthesis of function-described graphs from a set of graphs. Their paper [169] extended this work, describing distance measures and an efficient matching algorithm between attributed graphs and function-described graphs. Additionally, they employed function-described graphs for modeling and matching objects represented by attributed graphs. Cook *et al.* [38] introduced a structural search engine utilizing the hyperlink structure of the web and textual information to locate websites of interest. Hidovic and Pelilo proposed distance measures [92] for attributed graphs, based on the maximal similarity common subgraph of two graphs. This innovative approach facilitated effective matching between complex attributed graphs.

In 2022, Dwivedi *et al.* [66] proposed a novel technique to find similarity between two graphs using tree edit distance. To find the similarity between two graphs authors first convert them to spanning tree, then find the tree edit distance between these two spanning tree. Finally, graph similarity is computed by augmenting the tree edit distance with additional edit operations.

3.4.2 Continuous Optimization

Graph matching, at its core, presents a formidable challenge as it is fundamentally a discrete optimization problem. However, recent advancements have shown a promising approach that involves a transformation of this discrete problem into a continuous one. This innovative method aims to pave the way for the application of various optimization algorithms, ultimately culminating in a solution that can be translated back into the discrete domain.

An innovative class of error-tolerant graph matching techniques centers around the concept of relaxation labeling. This approach relies on the assignment of labels from a predetermined set to each node in the first graph, aligning them with nodes in the second graph. For each node in one graph, a vector of probabilities is generated based on a normal probability distribution. These probabilities govern the assignment of the node to potential corresponding nodes in the other graph. In the initial phase, these probabilities are computed utilizing node and edge attributes, as well as other pertinent graph-related information. Subsequently, through a series of iterative steps, these probabilities are refined until the desired labeling is achieved. Fischler and Elschlager introduced relaxation labeling [70]. Their work not only provided a combined descriptive scheme and decision metric but also indicated an algorithm that employed an approach akin to dynamic programming. This ingenious strategy significantly reduced computational time. Building upon this foundation, Rosenfeld *et al.* [157] extended the concept from discrete relaxation to probabilistic relaxation. This extension allowed for ambiguity reduction among objects in a scene through iterated relaxation operations.

Subsequent works witnessed significant enhancements and extensions to the relaxation labeling technique. Kittler and Hancock [108] sought to address the challenge of systematically combining evidence in probabilistic relaxation. Their work introduced a specification that eliminated the need for a support function. Building on this, the same authors further described discrete relaxation for performing maximum a posteriori probability estimation of globally consistent label assignments [88]. Christmas *et al.* [34] proposed probabilistic relaxation for graph matching, leveraging features extracted from 2D images and framing the matching problem within the Bayesian framework for label assignment. Rangarajan and Mjolsness [146] introduced a Lagrangian relaxation network tailored for graph matching, aiming to find a permutation matrix that aligns nodes from two graphs into a correspondence. Wilson and Hancock [194] expanded on this by proposing an error-tolerant matching technique using Bayesian consistency measures, extending the probabilistic framework. Huet and Hancock [97] extended this work further, presenting Bayesian matching for large image libraries, incorporating edge consistency alongside node attribute similarity.

Myers *et al.* [136] introduced Bayesian graph edit distance as a means of performing error-tolerant matching for corrupted graphs. Torsello and Hancock [183] utilized relaxation labeling to approximate tree edit distance, employing a uniform edit cost. Kostin *et al.* [113] extended this using the probabilistic relaxation algorithm proposed earlier [34]. Chevalier *et al.* [33] presented an enhancement of relaxation labeling techniques for region adjacency graphs, with the approach evolving over time. Finally, Sanroma *et al.* [162] put forth a graph matching technique leveraging the Expectation-Maximization algorithm to address the point-set correspondence problem. This innovative approach opened new avenues for solving intricate graph matching challenges. These advances collectively signify a significant leap forward in the field of graph matching, offering more robust and versatile solutions for a wide array of real-world applications.

3.4.3 Spectral Methods

The use of the spectral method, rooted in algebraic graph theory, constitutes a notable class of graph matching techniques. This approach capitalizes on the inherent stability of adjacency matrices, which remain unaltered even amongst node rearrangements. Consequently, for similar graphs, their adjacency matrices yield identical eigendecompositions. The computation of eigenvalues and eigenvectors, while well-known as challenging tasks, can be efficiently tackled within polynomial time. This attribute renders spectral techniques an attractive choice for various graph matching methodologies, given their capacity to elucidate crucial structural information through eigenvalues and eigenvectors.

Umeyama [188] introduced an innovative method for matching two weighted graphs, where each edge is assigned a specific weight, leveraging the eigendecomposition of adjacency matrices. Umeyama demonstrated that near-optimal matching outcomes could be achieved provided that the graphs exhibit a degree of proximity. However, it's worth noting that this method imposes a prerequisite: the input graphs must be of equal size. A seminal paper by Carcassoni and Hancock [29] presented an eigendecomposition-based technique tailored for weighted graph matching. This approach adopted a hierarchical strategy, employing a probabilistic framework to confine individual matching within pairwise clusters. Caelli and Kosinov [28] introduced a method based on graph eigen-space, facilitating error-tolerant graph matching. Their approach incorporated eigen-subspace projection and vertex clustering techniques, demonstrating applicability to graphs with varying node counts. Robles-Kelly and Hancock [154] proposed a graph spectral seriation technique that transforms the adjacency matrix of a graph into a string sequence. This transformation enabled the application of efficient string matching methods, expanding the potential applications of spectral techniques. Additionally, in the same year, Shokoufandeh *et al.* [176] outlined a framework for indexing hierarchical image structures, leveraging the spectral characterization of directed acyclic graphs. Cour *et al.* [46] advanced the field with their proposal of balanced graph matching, employing spectral relaxation techniques to yield approximate solutions for graph matching problem. Their work also introduced a normalization technique designed to heighten the precision of existing graph matching methodologies. Subsequently, Duchenne *et al.* [54] introduced a tensor-based algorithm for hypergraph matching. Their approach aimed to maximize the multilinear objective function by extending the principles of spectral techniques to hypergraphs, a significant leap in broadening the scope of spectral methods.

In summary, the spectral method's versatility and ability to filter essential structural information from graphs through eigenvalues and eigenvectors have led to its widespread adoption in graph matching techniques. Pioneering works by Umeyama, Carcassoni and Hancock, Caelli and Kosinov, Robles-Kelly and Hancock, Shokoufandeh *et al.*, Cour *et al.*, and Duchenne *et al.* have demonstrated the effectiveness and applicability of spectral techniques across various graph matching scenarios, solidifying its position as a crucial tool in this field.

3.4.4 Artificial Neural Networks

A fascinating category of graph matching techniques incorporates the use of artificial neural networks, providing a powerful computational framework for tackling complex matching problems. Suganthan *et al.* [182] pioneered the application of the Hopfield neural network to homomorphic graph matching, leveraging an input energy criterion for optimization. Their work marked a significant step towards harnessing the potential of neural networks in the realm of graph matching. A notable contribution by Sperduti and Starita [180], introduced a supervised neural network for graph classification. Their approach introduced the concept of generalized recursive structure, offering a novel perspective on utilizing neural networks for tasks involving graphs. This represented a key moment in the exploration of neural networks within the context of graph matching. Frasconi *et al.* [72] presented a pioneering framework that unified the processing of structural information by amalgamating artificial neural networks and belief network models. This innovative approach leveraged directed acyclic graphs to represent the connections between data variables, unlocking new directions for understanding and processing complex structural information. This work laid the foundation for subsequent developments in the intersection of neural networks and graph-based data.

Barsi [9] described the boundaries of self-organizing feature maps by introducing a novel variation. This extension replaced the conventional regular neuron grid with an undirected graph, guiding a fresh perspective on neural network architectures for graph-related tasks. This modification demonstrated the adaptability and versatility of neural networks in accommodating diverse data structures, including graphs. Jain and Wysotzki [102], proposed a neural network-based technique specifically tailored for error-tolerant graph matching problems. Their approach ingeniously reduced the search space by implementing a neural refinement procedure, followed by an energy minimization process. It demonstrated a subtle understanding of how neural networks can be utilized to address the intricacies of graph matching, particularly in scenarios where tolerance for errors is paramount. The paper by Scarselli *et al.* [165] introduced an innovative graph neural network model. This innovation extended existing neural network models to seamlessly handle data represented within the structural domain. By doing so, they addressed a critical gap in the field, enabling neural networks to directly process and extract meaningful insights from complex structural data, further expanding the applicability of this powerful computational paradigm.

In summation, the incorporation of artificial neural networks into graph matching methodologies represents a promising frontier in computational techniques. Pioneering works by Suganthan et al., Sperduti and Starita, Frasconi et al., Barsi, Jain and Wysotzki, and Scarselli et al. have collectively propelled the field forward, showcasing the versatility and efficacy of neural networks in handling complex graph-based data and underlining their potential for solving a wide array of practical problems.

3.4.5 Graph Edit Distance

Graph edit distance is a remarkably flexible and widely employed technique for performing error-tolerant graph matching, making it an indispensable tool in various domains. Fu and Bhargava [75] introduced a pioneering method that departed from the conventional representation of patterns as strings, opting instead for a tree-based approach. Their work involve the development of a comprehensive tree system encompassing grammar, transformation, and mapping on trees, along with tree automata. This innovative framework found practical application in the domain of structural pattern recognition, significantly broadening the scope of graph matching methodologies.

In a seminal paper, Sanfeliu and Fu [160] further advanced the field by presenting a novel distance measure for determining the minimum number of alterations required to transform one graph into another. This transformative measure operated on the principles of insertion, deletion, and substitution of both nodes and edges, offering a comprehensive approach to evaluating the structural dissimilarity between graphs. The introduction of these edit operations marked a significant step forward in the development of error-tolerant graph matching techniques, enhancing their adaptability to real-world applications. Building upon this foundation, Bunke and Allerman [24], contributed a pivotal advancement in error-tolerant graph matching of attributed graphs for structural pattern recognition. Their approach depended on the utilization of state space search, augmented with heuristic information, to guide the matching process. A defining feature of their work was the formulation of a judiciously designed cost function governing edit operations. This innovative cost function was carefully constructed to ensure that the resulting graph distance adhered to the essential properties of metrics under specific conditions. By imbuing the cost function with these fundamental properties, Bunke and Allerman significantly elevated the robustness and reliability of error-tolerant graph matching methodologies.

The applicability of graph edit distance spans a wide array of domains, owing to its flexibility to specific applications through the definition of adaptive edit cost functions. However, a notable limitation of this technique lies in its computational complexity, with a worst-case complexity that escalates exponentially with the number of nodes in the input graphs. This computational complexity was underscored by Zeng *et al.* [200], who formally demonstrated the NP-hardness of graph edit distance, emphasizing the need for more efficient solutions. In response to this computational limitation, a plethora of approximate and suboptimal algorithms for error-tolerant graph matching have emerged. Neuhaus and Bunke [137] introduced a method designed for planar graphs, leveraging a successive subgraph matching strategy to iteratively optimize structural similarity. This approach strategically refines the matching process by considering subgraphs constructed from the neighborhoods of nodes in both graphs. By iteratively pursuing this path, the algorithm incrementally refines the edit path, ultimately achieving a locally optimized solution.

Expanding on this foundation, Neuhaus *et al.* in [139], delved into techniques for suboptimal computation of graph edit distance. They identified that A^* search-based graph edit distance algorithms often explore an extensive search space, which may not be entirely relevant to specific classification problems. To address this, the authors proposed two variants of A^*: A^* beam search, which restricts exploration to a fixed number of nodes (beam-width) at each level of the search tree, and A* path length, which introduces an additional weight factor favoring longer partial edit paths over shorter ones.

Riesen and Bunke [149] proposed a suboptimal approach to graph edit distance computation, leveraging bipartite graph matching. This technique focused on local edge structures during optimization, as opposed to a global perspective. By employing bipartite optimization, the method mapped vertices and their local structures from the first graph to corresponding elements in the second graph. Sole-Ribalta *et al.* [177] provided a comprehensive overview of the properties and applications of graph edit distance cost, shedding light on its significance in various contexts. Ferrer *et al.* [69] enhanced the distance accuracy of the bipartite graph matching framework by strategically exploiting the order of assignments computed by the approximation algorithm. Building on this foundation, Riesen *et al.* in [153], outlined an approach for estimating exact graph edit distance through regression analysis, leveraging both lower and upper bounds derived from bipartite approximations. Meanwhile, Riesen and Bunke [151] explored diverse search strategies, such as iterative search, greedy search, and beam search, to further refine the approximation of bipartite graph edit distance computation.

In the realm of structural pattern recognition, Riesen in [147], described various algorithms employing graph edit distance. Dwivedi and Singh, in [60], introduced homeomorphic graph edit distance, which involved path contraction on the two input graphs before computing graph edit distance. Building on this work, in 2018, the same authors extended this approach to encompass error-tolerant graph matching, employing node contraction as a key element [62]. This iterative process of refinement and innovation in graph edit distance techniques reflects the field's ongoing commitment to overcoming computational challenges and advancing the applicability of this powerful matching methodology. In 2020, authors propose error-tolerant approximate graph matching utilizing node centrality information [64]. Inexact graph matching using various centrality measures is presented in [57]. In 2022, author describe algorithms to find approximate graph edit distance for structural pattern recognition [59]. Dwivedi and Singh in 2023 describe the recent techniques and algorithms to find graph edit distance and apply it to perform pattern recognition tasks [65].

3.4.6 Special Type of Graphs

Error-tolerant graph matching has been a subject of extensive research, leading to the development of specialized algorithms tailored to specific types of graphs. Lu [125], in a seminal work, introduced a groundbreaking algorithm for computing the optimal matching of two trees. This innovative approach relied on node splitting and merging techniques, significantly advancing the field of tree-based matching. Zhang *et al.* in [201] introduced a highly efficient polynomial-time algorithm designed specifically for calculating the editing distance between unordered labeled trees. Their contribution marked a crucial step forward in the development of algorithms capable of handling this particular type of graph. In [172], Shasha *et al.* made significant strides by presenting a series of efficient algorithms and heuristics, which led to approximate solutions for the tree edit distance problem. Their work not only demonstrated the feasibility of finding practical solutions but also paved the way for further advancements in this area.

Bernard *et al.*, in [12], delved into probabilistic models for tree edit distance, employing the expectation maximization algorithm to develop a robust framework for handling variations in tree structures. Oflazer, in publication, [143] introduced an error-tolerant tree matching algorithm utilizing a branch and bound technique. This approach strategically pruned the search space, enabling the algorithm to efficiently explore potential matches, particularly in cases where finding the exact matches might be challenging. Rocha and Pavlidis [155] presented a groundbreaking technique for error-correcting homomorphism tailored specifically for planar graphs. Their work addressed a critical need in the field, providing a specialized solution for a specific class of graphs. Valiente, in [189], introduced an algorithm for efficiently computing tree distance, offering a polynomial-time solution that significantly expanded the scope of feasible applications for this type of graph matching.

In a recent contribution, Dwivedi and Singh in [63] presented an efficient error-tolerant graph matching algorithm explicitly designed for geometric graphs. This work represents a cutting-edge advancement in the field, demonstrating the ongoing commitment to developing specialized solutions for various graph types. Moreover, the pursuit of inexact graph matching for specific classes of graphs has led to noteworthy contributions in works such as [191] and [124]. These works have further expanded the techniques for addressing the limitations and complexities inherent in graph matching for specialized graphs. Bipartite graph matching is one of the important technique for error-tolerant graph matching. In a recent work author in [58] present an approach to bipartite graph matching by considering the edge assignment in addition to node assignment during the creation of cost matrix.

3.4.7 Graph Kernels and Embeddings

Kernel methods have emerged as powerful tools for extending statistical pattern recognition techniques into the domain of graphs. The diverse array of graph kernels utilized in graph matching has been extensively documented by Neuhaus and Bunke [138] and Gartner [78], offering a comprehensive overview of the available techniques. Neuhaus *et al.* [140]

provided an in-depth exploration of various kernels specifically modified for error-tolerant graph classification. Their work extended the applicability of kernel methods to scenarios where robustness to errors and dissimilarities in graph structures is crucial. Kashima *et al.* in [105] introduced marginalized kernels designed for labeled graphs. Their approach harnessed the concept of an infinite-dimensional feature space, opening up new avenues for characterizing and comparing complex graph structures.

Another noteworthy contribution comes from [18], where the authors presented graph kernels based on the concept of shortest paths. Notably, these kernels possess the crucial property of positive definiteness and can be computed in polynomial time, making them practical for a wide range of applications. Lafferty and Lebanon in [116] introduced the diffusion kernel, a class of kernels designed for statistical learning. This kernel proved to be a valuable tool for tasks involving complex, interconnected data by leveraging the geometric structures inherent in statistical models. In a related work, [91] introduced convolution kernels specifically designed for discrete structures. This innovative approach, pioneered by Haussler, extended the class of radial basis kernels, enabling the characterization of intricate relationships within discrete data sets. Gazere *et al.* in [79] presented two novel graph kernels for regression and classification tasks. The Laplacian kernel, rooted in the concept of edit distance, and the treelet kernel, based on subtree enumeration, offered versatile tools for characterizing and comparing graphs.

Strug [181] introduced a kernel specifically designed for hierarchical graphs, leveraging a combination of tree and graph kernels. This approach facilitated the effective comparison of complex graph structures with hierarchical relationships. In [8], Bai and Hancock proposed graph kernels founded on the Jensen-Shannon kernel, a non-extensive information theoretic kernel. This innovative approach extended the applicability of information theoretic techniques to the domain of graph matching. Additionally, Riesen and Bunke [150] explored graph embedding techniques and their application to various classification and clustering tasks. By representing graphs in a vector space, these methods offered a powerful framework for characterizing and analyzing complex graph structures. This work represented a significant advancement in the field, opening up new avenues for graph-based data analysis and interpretation.

3.4.8 Other Methods

Now, we briefly survey other techniques for error-tolerant graph matching proposed in literature. In [127], a notable approach to error-tolerant graph matching is outlined, leveraging the power of expectation maximization algorithms. This method exploits the iterative process of expectation maximization to optimize graph matching, offering a robust framework for handling error tolerance. Gold and Rangarajan [82] introduced the graduated assignment algorithm. This algorithm offers an effective strategy for graph matching, providing a step-wise assignment process that gradually refines the matching solution. Its adaptability and efficiency have made it a valuable tool in a wide array of applications.

The work by Chang and Kimia [30] extended the graduated assignment algorithm to hypergraphs. This advancement allowed for the application of this technique to complex, high-dimensional data structures, opening up new possibilities for error-tolerant matching in a broader range of contexts. Liu *et al.* [122] proposed a modified genetic algorithm tailored specifically for solving the weighted graph matching problem. This approach leverages the evolutionary principles of genetic algorithms to search for optimal matching solutions in scenarios where weighted attributes play a crucial role. Khoo and Suganthan [107] introduced a genetic algorithm-based optimization technique for multiple relational graph mapping. This method optimizes the assignment of multiple graphs to a relational structure, offering a versatile tool for various applications, including data integration and knowledge representation. Wang *et al.* [192] presented an approximate graph matching algorithm designed for labeled graphs. This technique focuses on efficiently matching graphs with labeled nodes, accommodating scenarios where the semantic information associated with nodes is of paramount importance.

Mori *et al.* [135] emphasized the importance of global features in online character recognition in their work. By considering the holistic characteristics of handwritten characters, their approach provides a robust foundation for accurately recognizing and classifying characters in dynamic, real-time settings. In [195], Wu *et al.* introduced an object tracking benchmark alongside effective approaches for robust tracking. This contribution addressed the challenge of tracking objects in video sequences, offering state-of-the-art techniques for accurate and reliable object localization. Cilia *et al.* [36] put forth a feature ranking-based approach with a greedy search strategy in their paper, aimed at handwritten character recognition. This technique combines feature ranking and a systematic search process to identify the most discriminative features for accurate recognition of handwritten characters. Bouillon *et al.* in [20] presented a preprocessing enhancement for handwritten historical document analysis. This novel preprocessing step optimizes the input image, providing a cleaner and more structured basis for subsequent analysis and recognition tasks.

Finally, additional techniques for error-tolerant graph matching are described in [99, 101, 163, 179], demonstrating the diversity and richness of approaches in the field. These works contribute to a growing body of knowledge aimed at tackling the challenges posed by error-tolerant graph matching across various domains.

Brun *et al.* [21] provide recent trends into the evolving landscape of graph-based representation for pattern recognition. Their work provides an up-to-date perspective on how graphs are increasingly being employed as powerful tools for recognizing complex patterns in diverse domains. In [26], Bunke and Riesen conducted a comprehensive survey that sought to unify the realms of structural and statistical pattern recognition. By synthesizing these two pivotal areas, their survey explores the potential for synergistic approaches that leverage both structural and statistical information for pattern analysis. Hancock and Wilson [89] present a comprehensive review of the most recent advances in pattern analysis utilizing graph-based representations. This review encompasses a wide spectrum of techniques and methodologies, offering an extensive view of the cutting-edge developments in this dynamic field.

Jain *et al.* conducted an extensive survey in [100], providing a comprehensive overview of the biometric research landscape over the last five decades. Their work encapsulates the achievements, ongoing challenges, and future prospects in the domain of biometric pattern recognition, shedding light on the significant contributions made in this critical field. Santosh in [164] delves into the diverse array of statistical, structural, and syntactic approaches utilized in graphics recognition. This overview serves as a valuable resource for understanding the methodologies and techniques employed in recognizing and interpreting graphics across various applications and domains. Novel algorithms on exact, approximate, and error-tolerant graph matching are presented in [56]. This work introduces the recent graph matching techniques using the representational power of the graph and apply it to structural pattern recognition applications.

These surveys collectively offer a comprehensive view of the advancements, challenges, and opportunities in the realm of pattern recognition, particularly in the context of graph-based representation. They underscore the increasing significance of graphs as a versatile and powerful tool for pattern analysis across a wide spectrum of disciplines and applications.

3.5 Summary

In this chapter, a comprehensive foundation on graph matching is provided, covering fundamental concepts and definitions essential to this field. Moreover, we conducted an extensive survey of various graph matching techniques and algorithms, addressing both exact and error-tolerant graph matching methodologies.

For precise applications demanding rigorous matching criteria like graph isomorphism, exact graph matching proves to be indispensable. Conversely, in scenarios affected by noise or distortion, inexact matching emerges as the optimal solution. We have also described techniques for specialized graph types, catering to specific applications with distinct requirements.

During our survey of graph matching techniques, one prominent challenge that surfaced was the computational complexity associated with these methods. In subsequent chapters, specifically Chapters 4 and 5, present innovative approximate solutions to mitigate this issue. Chapter 4, introduce a node contraction approach, strategically reducing the search space based on node degree. Chapter 5 describe a novel matching technique derived from a subset of nodes in both graphs, employing centrality measures. This method proves invaluable for scenarios where time constraints play a pivotal role.

Yet another critical aspect we encountered in graph matching was the absence of a comprehensive similarity measure. Addressing this concern, Chapter 6 proposes a robust similarity measure for evaluating geometric graphs. This innovative framework requires that each node in the graph dataset be associated with specific coordinate points. By embedding any given graph in a two-dimensional plane, we can derive the coordinates of its vertices. Leveraging this framework, we achieve precise assessments for both exact and error-tolerant graph matching scenarios. This measure stands as a significant contribution to the field, enhancing the accuracy and effectiveness of graph matching methodologies.

Chapter 4

Graph Matching using Extensions to Graph Edit Distance

4.1 Introduction

Graph edit distance stands as a fundamental technique in graph matching, owing to its significance and wide-ranging applications. The basic principle underlying graph edit distance lies in to transforming one graph into another while incurring the minimal number of deformations. These transformations are encapsulated by edit operations, encompassing the insertion, deletion, and substitution of nodes and edges. The versatility of this distortion model, coupled with its associated cost, renders graph edit distance applicable across an extensive spectrum of scenarios.

A key strength of graph edit distance lies in its adaptability to graphs of arbitrary nature, be they labeled or unlabeled. Additionally, it excels in handling diverse forms of structural distortions, further enhancing its applicability. However, it is crucial to acknowledge the computational complexity associated with this technique. Specifically, the computation time for graph edit distance exhibits exponential growth in relation to the number of nodes within the input graphs.

This chapter describes the extensions of graph edit distance, strategically striking a balance between computation time and matching accuracy across various applications. This perspective to error-tolerant graph matching leverages the idea of graph homeomorphism to determine structural similarity between two graphs [60]. By utilizing path contraction, we effectively reduce the number of nodes in the input graphs, while maintaining their topological equivalence. Additionally, we introduce an alternative technique utilizing node contraction, wherein we transform a given graph into another by consolidating nodes with smaller degrees [62]. Both strategies culminate in a notable reduction of the requisite search space for computing graph edit distances between the transformed graphs.

This chapter is structured as follows: Section 4.2 provides an in-depth exploration of the preliminary concepts, while Section 4.3 unveils the intricacies of homeomorphic graph edit distance. Section 4.4 introduces the concept of error-tolerant graph matching using extension to graph edit distance. The practical performance of these methods is demonstrated in Section 4.5.

DOI: 10.1201/9781003532248-4

Finally, Section 4.6 offers a concise summary of this chapter, encapsulating its key insights and contributions.

4.2 Preliminaries

This section describes the elementary definitions and notations behind graph matching, laying the groundwork for our proposed methodology.

Let us consider two graphs G_1 and G_2. A bijective mapping denoted as $f : V_1' \rightarrow V_2'$ from G_1 to G_2 is said to be an *error-tolerant graph matching*, if $V_1' \subseteq V_1$ and $V_2' \subseteq V_2$ [22].

Adding nodes or edges, removing nodes or edges, and substituting nodes or edges are some of the frequently performed edit operations in the field of graph matching. These operations are symbolically represented as follows: Insertion of a vertex u is denoted by $\epsilon \rightarrow u$, while the deletion of vertex u is represented by $u \rightarrow \epsilon$, and substitution, a more intricate operation, involves the replacement of a vertex u by vertex v is denoted by $u \rightarrow v$. Additionally, we include an edit operation for path contraction, enabling the contraction of a simple path wherein every vertex, except the first and last, has a degree of two. This transformation is ably represented as $(u_1, ..., u_n) \rightarrow (u_1, u_n)$. By employing path contraction judiciously, we effectively eliminate nodes with a degree of two, while upholding the topological equivalence of the graphs.

Consider $G = (V, E)$ be a graph with an edge $e = (u, v) \in E$. Subdivision of the edge e generates another graph $G' = (V', E')$ such that $V' = V \cup \{w\}$, and $E' = E \setminus \{(u, v)\} \cup \{(u, w), (w, v)\}$, here w is an additional vertex and (u, w), (w, v) are additional edges. A subdivision of a graph G produces another graph G' by performing the subdivision operation on some edges in G. Two graph G_1 and G_2 are said to be homeomorphic or topologically equivalent, if both graphs G_1 and G_2 are a subdivision of some graph G.

A noteworthy observation emerges from the subdivision operation: it exclusively impacts the count of vertices with a degree of two. Consequently, for two graphs to be deemed homeomorphic, they must share an identical count of vertices for every degree except two. Two graphs must have the same number of vertices in every degree except two, in order to be homeomorphic, as subdivision of G only modifies the number of vertices of degree two.

4.3 Homeomorphic Graph Edit Distance

Two graphs are called homeomorphic when one of the graphs can be transformed into the other through subdivision operations on their edges, accomplished by inserting additional nodes along these edges.

Consider a graph $G = (V, E)$ with V as the vertex set and E as the edge set. Let $e = (u, v) \in E$. The subdivision operation on the edge E produces another graph $G' = (V', E')$ with $V' = V \cup \{w\}$, and $E' = E \backslash \{(u, v)\} \cup \{(u, w), (w, v)\}$, introducing a new vertex w and connecting it with (u, w) and (w, v) edges.

The smallest number of edit operations needed to transform one graph, G_1 into a homeomorphic form with respect to another graph, G_2 is known as the homeomorphic graph edit distance. The fundamental idea is that two graphs, with the exception of degree two vertices, must have the same number of vertices for each degree in order to be homeomorphic or topologically identical. Consequently, we can eliminate vertices with a degree of two, resulting in a modified graph that remains homeomorphic.

Definition 4.1 *Let $G_1 = (V_1, E_1, \mu_1, \nu_1)$ and $G_2 = (V_2, E_2, \mu_2, \nu_2)$ be two graphs. The homeomorphic graph edit distance (HGED) between G_1 and G_2 is defined as*

$$HGED(G_1, G_2) = GED(G'_1, G'_2)$$

$$= min_{(e_1,\ldots,e_k) \in \varphi(G'_1, G'_2)} \sum_{i=1}^{k} c(e_i) \tag{4.1}$$

Here G'_1 and G'_2 are graphs derived from G_1 and G_2 respectively by executing path contraction from each vertex. $\varphi(G'_1, G'_2)$ represents the set of edit distance paths transforming G'_1 into G'_2, and $c(e_i)$ denotes the cost function corresponding to each edit operation e_i.

Specifically, path contraction is performed on all simple paths $(u_1, ..., u_n)$ in a graph, ensuring that all vertices along this path have a degree of two (for $i = 2$ to $n - 1$, $Deg(u_i) = 2$) excluding first (u_1), and last vertex (u_n). These path contractions significantly reduce the processing load involving numerous vertices and edges, which is otherwise necessary for computing the graph edit distance.

Proposition 4.1 *Consider a graph $G = (V, E, \mu, \nu)$. If we obtain another graph $G' = (V', E', \mu', \nu')$ by performing path contraction $(u_1, ..., u_n) \rightarrow (u_1, u_n)$, effectively eliminating or smoothing the vertices $u_2, ..., u_{(n-1)}$ where $Deg(u_i) = 2$ for $i = 2$ to $n - 1$, then G and G' are homeomorphic.*

Nodes with a degree of two are specifically removed during the path contraction procedure, leaving the remaining nodes intact. It follows that G and G' are homeomorphic since smoothing reverses the subdivision operation.

4.3.1 Homeomorphic Edit Cost

Homeomorphic edit cost functions rely on the Euclidean distance measure to quantify the similarity between graphs. In this context, a constant cost is assigned to the operations of inserting and deleting both nodes and edges. For substitution operations involving nodes, edges, and paths, the cost is determined proportionally to their respective Euclidean distances.

This approach ensures that the cost associated with modifying the structure of graphs is consistent and reflective of the spatial relationships between their components. The Euclidean distance, a fundamental geometric measure, serves as a metric to gauge the dissimilarity or alteration between the elements of the graphs, enabling a comprehensive evaluation of the homeomorphic edit cost.

Let G_1 and G_2 be two graphs, we define homeomorphic graph edit cost function for all nodes $u \in V_1$, $v \in V_2$ and for all edges $e \in E_1$, $e' \in E_2$ by

$$
\begin{aligned}
&c(u \rightarrow \epsilon) = x_{node} \\
&c(\epsilon \rightarrow v) = x_{node} \\
&c(u \rightarrow v) = y_{node}.||\mu_1(u) - \mu_2(v)|| \\
&c(e \rightarrow \epsilon) = x_{edge} \\
&c(\epsilon \rightarrow e') = x_{edge} \\
&c(e \rightarrow e') = y_{edge}.||\nu_1(e) - \nu_2(e')|| \\
&c((u_1, ..., u_n) \rightarrow (u_1, u_n)) = z_{path}.||\mu_1(u_1) - \mu_1(u_n)||
\end{aligned}
$$

In this formulation x_{node}, x_{edge}, y_{node}, y_{edge}, and z_{path} are non-negative values. The terms $c(u \rightarrow \epsilon)$, $c(\epsilon \rightarrow v)$, $c(u \rightarrow v)$, $c(e \rightarrow \epsilon)$, $c(\epsilon \rightarrow e')$ and $c((u_1, ..., u_n) \rightarrow (u_1, u_n))$ represent the cost associated with deleting a node u, inserting a node v, substituting node u with node v, deleting an edge e, inserting an edge e', and path contraction from $(u_1, ..., u_n)$ to (u_1, u_n) respectively.

It is remarkable that the substitution cost, as defined, is proportional to the Euclidean metric of the corresponding vertex and edge labels. This choice of metric aligns with the geometric relationships between the labels, contributing to a comprehensive assessment of the dissimilarity between the graphs.

This graph edit function bears similarities to the edit cost used in graph edit distance, with the notable addition of the path contraction operation $(u_1, ..., u_n) \rightarrow (u_1, u_n)$ and the introduction of the parameter z_{path}.

Proposition 4.2 *The substitution of a path, denoted as* $c((u_1, ..., u_n) \rightarrow (u_1, u_n))$ *can results in savings of up to* $(2n - 2)$ *edit operations.*

In this context, it is evident that the parameter z_{path} can be expressed as $z_{path} = (n-1)x_{edge} + (n-2)x_{node} + x_{edge} = n.x_{edge} + (n-2)x_{node}$. Consequently, this formulation allows for the potential saving of $(2n-2)$ operations for each path contraction.

To illustrate, consider the computation of a homeomorphic edit path from graph G to graph H, as depicted in Figure 4.1. Initially, G is transformed into G' and H is transformed into H' using path contraction. Afterwards, four edit operations are performed to convert G' to H': node insertion, edge insertion, and node insertion again, followed by another edge insertion.

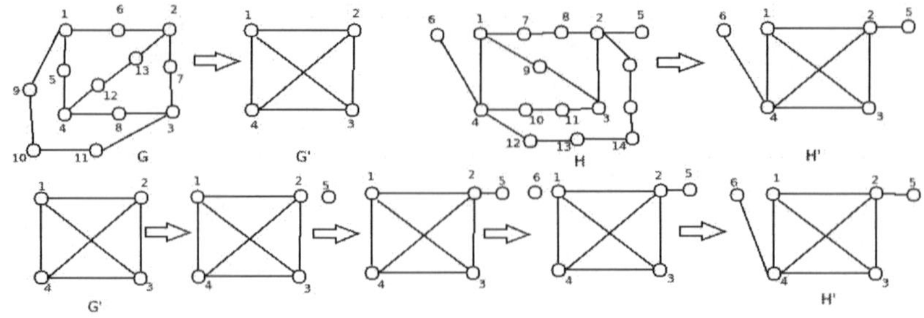

FIGURE 4.1: Homeomorphic edit path from G to H

4.3.2 Algorithm

Algorithm 6 outlines the procedure for computing error-tolerant graph matching through the utilization of graph homeomorphism. The Homeomorphic-Graph-Edit-Distance algorithm takes two input graphs, $G_1 = (V_1, E_1, \mu_1, \nu_1)$ and $G_2 = (V_2, E_2, \mu_2, \nu_2)$. Graph G_1 contains n nodes whereas graph G_2 consists of m nodes. The output of the algorithm is the least cost homeomorphic graph edit distance from G_1 to G_2

The algorithm's steps involve path contraction on both G_1 and G_2. For each vertex in G_1, the algorithm searches for simple paths where all intermediate vertices have a degree of two. Subsequently, it updates these paths by removing the intermediate vertices and inserting an edge between the first and last vertices. The same path contraction operations are then performed on G_2, updating both graphs and their parameters.

The algorithm initializes an empty set A and conducts simultaneous substitution of the first vertex, u_1 of G_1 with every other vertex of G_2, adding the deletion of u_1 to A. A while loop is run until a minimum cost edit distance, C_{min}, is found, which represents a complete graph edit path. Pruning on set A is done inside the loop via a variety of heuristic and optimization techniques to narrow down the search space.

The algorithm outputs C_{min} if it is the complete edit path, meaning that the set of graph edit paths is the same as the initial graph. However, if all nodes of G_1 are processed and some nodes of G_2 remain unprocessed, these unprocessed nodes are added to C_{min}, and set A is updated in the inner loop of the if statement. All of the nodes from the second graph, G_2, are used to replace each unprocessed node of G_1 in the else statement's inner loop. Set A has been modified as a result of these substitutions being added to C_{min} and the removal of G_1's unprocessed nodes.

The validity of the Homeomorphic-Graph-Edit-Distance algorithm can be verified through Proposition 4.1. This algorithm achieves a reduction in the number of vertices within the input graph, thereby diminishing the search space needed for subsequent edit distance computations. To further optimize the execution time, several pruning strategies can be employed. In set A, for

Algorithm 6 : Homeomorphic-Graph-Edit-Distance (G_1, G_2)

Require: Two Graphs G_1, G_2, where $G_i = (V_i, E_i, \mu_i, \nu_i)$ for $i = 1, 2$ where $V_1 = \{u_1, ..., u_n\}$ and $V_2 = \{v_1, ..., v_m\}$
Ensure: A min. cost homeomorphic GED between G_1 and G_2

for each $(u_i \in V_1)$ **do**
    if (there is a path $(u_i, u_{i+1}, ..., u_{i+k})$ such that $deg(u_{i+1}) = deg(u_{i+2}) = ... = deg(u_{i+k-1}) = 2$) **then**
        $(u_i, u_{i+1}, ..., u_{i+k}) \rightarrow (u_i, u_{i+k})$
        $V_1 \leftarrow V_1 \setminus \{u_{i+1}, ..., u_{i+k-1}\}$
    end if
end for
for each $(v_j \in V_2)$ **do**
    if (there is a path $(v_j, v_{j+1}, ..., v_{j+k})$ such that $deg(v_{j+1}) = deg(v_{j+2}) = ... = deg(v_{j+k-1}) = 2$) **then**
        $(v_j, v_{j+1}, ..., v_{j+k}) \rightarrow (v_j, v_{j+k})$
        $V_2 \leftarrow V_2 \setminus \{v_{j+1}, ..., v_{j+k-1}\}$
    end if
end for
Update $G_1, G_2, n \leftarrow n', m \leftarrow m'$
$A \leftarrow \emptyset$
for each $(v_j \in V_2)$ **do**
    $A \leftarrow A \cup \{u_1 \rightarrow v_j\}$
end for
$A \leftarrow A \cup \{u_1 \rightarrow \epsilon\}$
while (1) **do**
    Prune A using optimizing techniques
    Compute min. cost graph edit path C_{min} from A
    if (C_{min} is a complete graph edit path) **then return** C_{min}
    else
        if (all nodes $(u_i \in V_1)$ are processed) **then**
            for each unprocessed $(v_j \in V_2)$ **do**
                $C_{min} \leftarrow C_{min} \cup \{\epsilon \rightarrow v_j\}$
            end for
            $A \leftarrow A \cup \{C_{min}\}$
        else
            for (each unprocessed node $(u_i \in V_1)$) **do**
                for (each $(v_j \in V_2)$) **do**
                    $C_{min} \leftarrow C_{min} \cup \{u_i \rightarrow v_j\} \cup \{u_i \rightarrow \epsilon\}$
                end for
            end for
            $A \leftarrow A \cup \{C_{min}\}$
        end if
    end if
end while

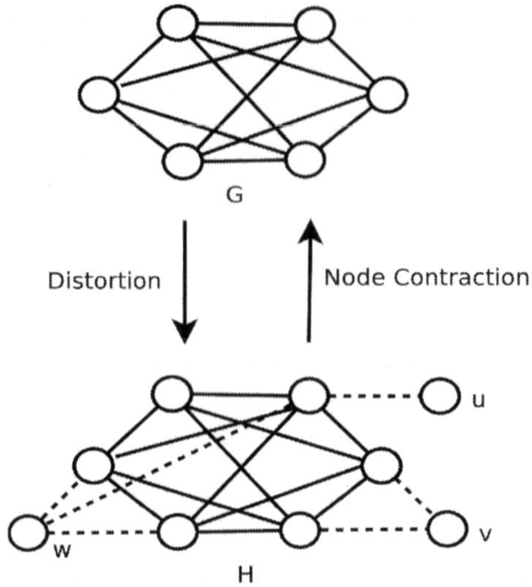

FIGURE 4.2: Node contraction on the graph H

example, a specific number of vertices can always be maintained. Furthermore, out of all the edit path entries that are accessible, only the minimal cost fixed entries are kept when there is an update in A. This method helps to make the computation process more streamlined and effective.

4.4 Extended Graph Edit Distance

Error-tolerant graph matching is important because it can correct faults that get incorporated within the graph being used while analyzing and retrieving as a result of interference and distortion. In the case of a dense graph, where distortions involving the adding or deleting of the nodes or edges are followed by subsequent adding or deleting, it might be practical to overlook the nodes with lowest degree when aiming to achieve graph matching with the minimal number of nodes. Consider a scenario where a graph G is initially presented, as illustrated in Figure 4.2. Due to distortions and noise, additional nodes (u, v, w) are introduced, connected by dotted edges in graph H. By eliminating or contracting the nodes of degree 1, 2, and 3 the resulting graph remains identical to G, showcasing the resilience of error-tolerant graph matching to such distortions.

The concept behind node contraction is to disregard less significant nodes in a graph before performing in graph matching. Degree centrality serves as one of the metrics for node importance within a graph. Using degree centrality as an indicator, nodes with lower degrees are ignored or deleted, starting from the least significant ones. However, simply removing nodes can significantly alter the graph's topology. To strike a balance where the graph's topology undergoes gradual changes, nodes and their corresponding edges are deleted only if this action does not disconnect the graph. Essentially, we eliminate nodes having lower degrees such that the graph's overall count of connected components stays fixed. Considering this viewpoint, we can extend the graph edit distance to avoid lower-degree nodes.

Definition 4.2 *Node contraction is the process of deleting a node and its corresponding edges provided that it is not a cut-vertex.*

A cut-vertex, sometimes called a cut-point or articulation point, is a vertex that when removed causes the graph to become disconnected.

Definition 4.3 *The technique of carrying out node contraction on every node in a graph G having a degree of k is known as k-degree node contraction. To denote k-degree node contraction on a graph G, we consider the notation $kNC(G)$.*

Definition 4.4 *Performing k-degree node contractions, which range from one degree to k, one after the other on a graph G represents k^*-degree node contractions.*

k^*-$NC(G)$=$kNC((k-1)NC((k-2)NC(...1NC(G)...)))$, where k^*-$NC(G)$ is k^*-degree node contraction on a graph G.

Definition 4.5 *By using k-degree node contractions on both graphs, k-GED calculates the graph edit distance between two graphs.*

Let G_1 and G_2 be two graphs, we define

$$k\text{-}GED(G_1, G_2) = GED(kNC(G_1), kNC(G_2))$$

When $k = 2$, k-GED generalizes to HGED [60], where path contraction is done such that every simple paths $(u_1, ..., u_n)$ are replaced by (u_1, u_n) where $Deg(u_i) = 2$ for $i = 2$ to $n - 1$.

Definition 4.6 *k-degree node contraction across both graphs, beginning at one-degree nodes and continuing to k-degree nodes, is used to compute the graph edit distance between two graphs in k^*-GED.*

For two graphs G_1 and G_2, we define

$$k^*\text{-}GED(G_1, G_2) = \begin{cases} GED(G_1, G_2), \text{if } k = 0 \\ GED(k^*\text{-}NC(G_1), k^*\text{-}NC(G_2)), \text{if } k \geq 1 \end{cases}$$

To provide a more specific characterization of the impact of node contraction on graph topology, we introduce the concept of node deletion. The term

"*Node deletion*" describes deleting a node along with all of its connected edges, regardless of whether or not it is a cut-vertex. We can describe k-degree node deletion,k^*-ddegree node deletion (k^*-ND), k-ND-GED, and k^*-ND-GED, respectively, in a manner similar to k-degree node contraction, k^*degree node contraction (k^*-NC), k-GED, and k^*-GED. In the above definitions, the node contraction operation is replaced by the node deletion operation. It should be noted that $|k^*\text{-ND}(G)| \leq |k^*\text{-NC}(G)|$, where $|G|$ indicates the number of nodes in graph G. As the node deletion operation is unconstrained, the number of nodes in a graph G after node deletion is fewer than or same to the number of nodes after node contraction.

4.4.1 Extended Edit Cost

To establish the edit cost of k-GED, we can expand upon the edit cost of graph edit distance by introducing the term $c(u \to \epsilon) = 0$, specifically when $deg(u) = k$ and u is not a cut vertex in the graph.

k-GED uses the Euclidean distance measure and requires a fixed cost for adding, removing, and substituting vertices and edges. Here is how we construct the extended edit cost function, given two graphs, G_1 and G_2, and all nodes $u \in V_1$, $v \in V_2$, and all edges $e \in E_1$, $f \in E_2$.

$$
\begin{aligned}
&c(u \to \epsilon) = x_{node} \\
&c(\epsilon \to v) = x_{node} \\
&c(u \to v) = y_{node} \cdot ||\mu_1(u) - \mu_2(v)|| \\
&c(e \to \epsilon) = x_{edge} \\
&c(\epsilon \to f) = x_{edge} \\
&c(e \to f) = y_{edge} \cdot ||\nu_1(e) - \nu_2(f)|| \\
&c(u \to \epsilon) = 0, \text{ if } deg(u) = k \text{ and } u \text{ is not a cut point}
\end{aligned}
$$

Where, $c(u \to \epsilon)$ represents the cost of deleting a node u, $c(\epsilon \to v)$ denotes the cost of inserting a node v, $c(u \to v)$ is the cost of substituting a node u by node v, $c(e \to \epsilon)$ stands for the cost of deleting an edge e, $c(\epsilon \to f)$ is the cost of inserting an edge f, and x_{node}, y_{node}, x_{edge}, y_{edge} are non-negative constants.

We observe that the cost functions stated above satisfy the non-negativity property, i.e., $c(e_i) \geq 0$, for each node and edge edit operation e_i, while also satisfying the triangle inequality property for the inserting, deleting, and substituting the nodes and edges.

4.4.2 Algorithm

This section presents the algorithm that uses node contraction to conduct error-tolerant graph matching. Algorithm 7 describes the computation of k^*-degree node contraction of an input graph. The graph G and the parameter k are the inputs of the k^*-Node-Contraction algorithm. Lines 1–16 of the algorithm's outer *for* loop conduct k-degree node contraction iteratively from

Algorithm 7 : k^*-Node-Contraction (G, k)

Require: A graph $G = (V, E, \mu, \nu)$ and a parameter k

Ensure: Transformed graph after applying k^*-degree node contraction on G

```
1: for (i ← 1 to k) do
2:     for each (u ∈ G) do
3:         if ((deg(u)==k)) then
4:             u.visit ← 1
5:         else
6:             u.visit ← 0
7:         end if
8:     end for
9:     for each (u ∈ G) do
10:        if ((u.visit==1)&&(u is not cut vertex)) then
11:            V ← V \ {u}
12:            E ← E \ {(u, v)|(u, v) ∈ E, ∀v ∈ G}
13:            u.visit ← 0
14:        end if
15:    end for
16: end for
17: return G
```

1 to k. A boolean flag *visit* is used in the *for* loop of the algorithm in lines 2–8. It is reset to 0 otherwise and set to 1 when a vertex of G has a degree of k. During the *for* loop of lines 9—15, the visit flag gets changed to 0 and the node and its related edges are deleted from G if its visit flag is set to 1 and it is not a cut vertex. Finally, line 17 returns the altered graph G. The suggested error-tolerant graph matching framework uses this approach as a preprocessing step.

Algorithm 8 outlines the computation process for k^*-GED between two graphs. The k^*-Graph-Edit-Distance algorithm takes as input two graphs, G_1, G_2, along with a parameter k, and outputs the minimum cost k^*-GED between G_1 and G_2. It employs Algorithm 7 to perform k^*-degree node contraction on G_1 and G_2 in lines 1–2, resulting in transformed graphs G'_1 and G'_2 with vertex sets $V'_1 = \{u'_1, ..., u'_{n'}\}$ and $V'_2 = \{v'_1, ..., v'_{m'}\}$, respectively. The algorithm initializes an empty set S in line 3. In the *for* loop spanning lines 4–6, S is updated by substituting vertex u'_1 of G'_1 with each vertex of G'_2, and the deletion of u'_1 in line 7 is added to S. The *while* loop in lines 8–28 is utilized to compute the minimum cost edit path C_{min} from S.

A tree-based search algorithm, such as A^* search, is typically used to compute the lowest-cost edit path. In this algorithm, the foremost node, or root, indicates the initial edit operation, and the leaf node indicates the last edit operation. The edit path from the root to leaf nodes forms a complete edit distance path, illustrating the list of edit operations required to convert one graph into another, though this path may or may not be the optimum. To find

Algorithm 8 : k^*-Graph-Edit-Distance (G_1, G_2)

Require: Two Graphs G_1, G_2, where $G_i = (V_i, E_i, \mu_i, \nu_i)$ for $i = 1, 2$ where $V_1 = \{u_1, ..., u_n\}$ and $V_2 = \{v_1, ..., v_m\}$ and a parameter k
Ensure: A min. cost k^*-GED between G_1 and G_2
1: $G_1' \leftarrow k^*$-Node-Contraction(G_1, k) // $V_1' = \{u_1', ..., u_{n'}'\}$
2: $G_2' \leftarrow k^*$-Node-Contraction(G_2, k) // $V_2' = \{v_1', ..., v_{m'}'\}$
3: $S \leftarrow \emptyset$
4: **for** all $(v_i' \in V_2')$ **do**
5: $S \leftarrow S \cup \{u_1' \rightarrow v_i'\}$
6: **end for**
7: $S \leftarrow S \cup \{u_1' \rightarrow \epsilon\}$
8: **while** $(True)$ **do**
9: Prune S using heuristic methods
10: Compute C_{min} the min. cost edit path from S
11: **if** (C_{min} is a complete edit distance path) **then**
12: **return** C_{min}
13: **else**
14: **if** (every node $(u_i' \in V_1')$ is processed) **then**
15: **for** every unprocessed $(v_j' \in V_2')$ **do**
16: $C_{min} \leftarrow C_{min} \cup \{\epsilon \rightarrow v_j'\}$
17: **end for**
18: $S \leftarrow S \cup \{C_{min}\}$
19: **else**
20: **for** (each unprocessed node $(u_i' \in V_1')$) **do**
21: **for** (every $(v_j' \in V_2')$) **do**
22: $C_{min} \leftarrow C_{min} \cup \{u_i' \rightarrow v_j'\} \cup \{u_i' \rightarrow \epsilon\}$
23: **end for**
24: **end for**
25: $S \leftarrow S \cup \{C_{min}\}$
26: **end if**
27: **end if**
28: **end while**

the optimal edit path, every level of the search must take into account every arrangement of edit operations in the set S, which can be computationally costly. An alternate strategy is to use different heuristic methods to prune the set S (line 9), taking into account only a portion of the edit operations, which results in a suboptimal but still quite efficient solution. Beam search, for example, can be used to restrict the search space by concentrating on the top w best candidates at each search level, where w is the beam width. In line 12, the algorithm returns its result if C_{min} reflects the entire edit distance path throughout algorithm execution. If not, lines 14–18 update S and all unprocessed nodes of G_2' are added to C_{min}. The unprocessed vertices of G_1' are then deleted in lines 20–24 and every unprocessed node of G_1' is replaced with a node of G'_2. These actions are then added to C_{min}. Line 25 then makes an update to S.

4.4.3 Computational Analysis

Proposition 4.3 *k^*-Node-Contraction algorithm calculates k^*-degree node contraction of G_1 and G_2.*

We can note that the *for* loop in lines 1–16 of Algorithm 7 ensures that k-degree node contraction is initiated from i=1 to k. The *if* loop in line 10 permits the removal of nodes only if their degree is k and they are not cut vertices; however, the *visit* flag makes sure that every node is only taken into account for contraction once.

Proposition 4.4 *k^*-Node-Contraction algorithm runs in $O(n)$ time.*

The verification of whether a node is a cut vertex can be conducted in $O(n)$ time. Therefore, the *for* loop in lines 9–15 takes $O(n)$. Similarly, the *for* loop in lines 2–8 also takes $O(n)$ time. Considering the outer *for* loop in lines 1–16 iterates k times, the overall time complexity of Algorithm 7 is $O(k.n)$, which simplifies to $O(n)$ time.

Proposition 4.5 *k^*-Graph-Edit-Distance algorithm computes error-tolerant graph matching of G'_1 and G'_2.*

It emanates from the characteristics of edit operations, including the addition, deletion, and substitution of nodes and edges, combined with the definition of k^*-GED. The completeness of the edit path returned by Algorithm 8 guarantees a unique mapping for each node of G'_1 to G'_2. Simultaneously, Algorithm 7 ensures that the modified vertex sets, V'_1 and V'_2, are subsets of V_1 and V_2, respectively.

The worst-case complexity of the k^*-Graph-Edit-Distance algorithm exhibits exponential growth with respect to the number of nodes in the input graphs. However, selecting a suitable parameter k can mitigate the overall execution time. Let's consider an implementation of the k^*-Graph-Edit-Distance algorithm using a tree-search based method. At the first level of execution, u'_1 is replaced by each of the v'_i for $i = 1$ to m', resulting in a time complexity of $O(m')$. Suppose the substitution of u'_1 to v'_1 is chosen at the first level; then, during the next level, u'_2 is again replaced by each of the v'_i for $i = 2$ to m', incurring a time complexity of $O(m')$, leading to a total time complexity of $O(m'^2)$ up to the second level. Similarly, at the n'th level, the worst-case execution time becomes $O(m'^{n'})$.

It is clear that the topologies as well as the size of the input graphs have a considerable impact on the values of n' and m', making it necessary to evaluate the effect of k on the sizes of the transformed graphs n' and m'. Suppose we have a random graph $G(n, p)$, where n is the number of nodes and p is the probability that there are edges connecting every pair of vertices. The vertex degree distribution, denoted by p_k, follows a loose relationship between n and n', given by $p_k = \binom{n-1}{k} p^k (1-p)^{n-k-1}$.

Proposition 4.6 *In the context of a random graph $G(n, p)$, the values of n and n' utilized in Algorithm 8 adhere to the following inequality.*

$$n - \sum_{n=1}^{k} \binom{n-1}{k} p^k (1-p)^{n-k-1} \leq n' \leq n - (n-1)p(1-p)^{n-2}.$$

Proof *Here, n represents the number of nodes in G_1 and n' is the number of nodes in G'_1 obtained after k^*-NC(G_1). Consequently, the count of contracted nodes is given by $n - n'$. We observe that the number of nodes removed during k^*-NC(G_1) $\leq n - n' \leq$ number of nodes removed during k^*-ND(G_1), which implies $(n-1)p(1-p)^{n-2} \leq n - n' \leq \sum_{n=1}^{k} \binom{n-1}{k} p^k (1-p)^{n-k-1}$, where left most term is the number of nodes removed during 1^*-NC or 1^*-ND, since both are equivalent operations. Now multiplying the above expression by -1 and rearranging the terms $-\sum_{n=1}^{k} \binom{n-1}{k} p^k (1-p)^{n-k-1} \leq n' - n \leq -(n-1)p(1-p)^{n-2}$, by adding n in above expression we get $n - \sum_{n=1}^{k} \binom{n-1}{k} p^k (1-p)^{n-k-1} \leq n' \leq n - (n-1)p(1-p)^{n-2}$.*

4.5 Results and Discussion

4.5.1 Homeomorphic Graph Edit Distance

The Homeomorphic-Graph-Edit-Distance algorithm was applied in our studies on random graphs with a given number of nodes. Figure 4.3 shows a comparison of the execution timings for graph edit distance computation using a homeomorphic approach and graph edit distance computation using an unoptimized or pruned method. The mean value obtained from several runs of the homeomorphic and graph edit distance computations on different random graphs with a fixed number of nodes is used to calculate each execution time that is displayed. Interestingly, beyond 10 nodes, simple graph edit distance computation frequently failed to finish. On the other hand, homeomorphic graph edit distance computation showed that it could be executed successfully even for larger sizes.

In essence, homeomorphism is a topological notion that describes the topological resemblance between two objects or spaces. This notion can be extended to evaluate the structural resemblance of graphs. A set of homeomorphic graphs exhibits a certain level of resemblance, which is measured by the homeomorphic graph edit distance. Graphs that are homeomorphic to one another will have the same homeomorphic graph edit distance. For example, the homeomorphic graphs shown in Figure 4.4 have the same homeomorphic graph edit distance.

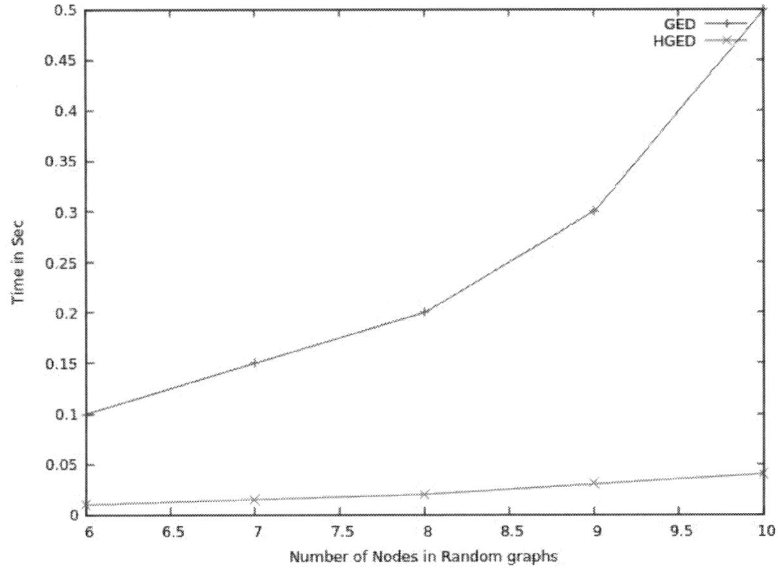

FIGURE 4.3: GED vs HGED computation

This concept proves especially beneficial in real-world applications where data may undergo modifications due to noise and distortion. The Homeomorphic-Graph-Edit-Distance algorithm can be employed for error-tolerant graph matching, offering versatility in applications such as pattern recognition, biometric identification, image processing, as well as in biological and chemical domains.

The suggested homeomorphic graph edit distance can be viewed as a 2-degree path contraction, achieved by eliminating all vertices along a path with a degree of 2. This concept can be extended to n-degree vertex path contraction, wherein a given graph transforms into another graph by removing all vertices with degrees up to n along any simple path in the input graph. The selection of the parameter n can be tailored to the specific requirements of the application.

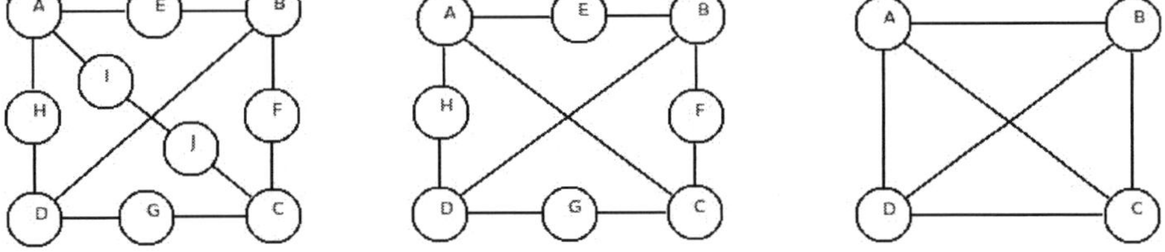

FIGURE 4.4: Homeomorphic Graphs

4.5.2 Extended Graph Edit Distance

We compare the suggested graph matching framework to other important graph matching techniques in this section. We use the graph matching toolkit [152] and the IAM graph database [148] to evaluate the accuracy and execution time of k^*-Graph-Edit-Distance. The IAM graph repository's letter and AIDS databases are used to generate the results.

4.5.2.1 Dataset Description

Letter graphs are made entirely of straight lines and contain all of the alphabet's uppercase letters. Fifteen classes of capital letters are included in the set: A, E, F, H, J, K, L, M, N, T, V, W, X, Y, and Z. Each prototype graph is subjected to three different levels of distortion (high, medium, and low) to produce a variety of graph datasets. The dataset for letters shows a uniform distribution among the fifteen letters, with nodes labeled with (x, y) coordinates that indicate where they are in relation to a reference plane. The average number of nodes and edges per graph in the letter graph data for the high-distortion level is 4.7 nodes and 4.5 edges, respectively. In this dataset, the maximum number of nodes and edges per graph is nine.

The AIDS dataset consists of graphs that classify chemical substances into confirmed active and confirmed inactive groups based on their characteristics. While inactive class molecules show inactivity against HIV, active class molecules show activity against the virus. Chemical symbols are represented by node labels, while valence is shown by edge labels. In the AIDS dataset, there are 15.7 nodes on average and 16.2 edges on average per network. This dataset allows for a maximum of 95 nodes and 103 edges per graph, respectively. A synopsis of the letter and AIDS dataset descriptions may be found in Table 4.1. The percentage of graphs with respect to size for both active and inactive molecules in the AIDS dataset training set is shown in Figure 4.5. It has been noted that active molecules typically have a larger number of nodes than inactive ones.

TABLE 4.1: Summary of letter and AIDS datasets

Description	Letter Dataset	AIDS Dataset
Patterns	Letter line drawings	Chemical compounds
Node labels	(x, y) coordinates	Chemical symbol
Edge labels	none	none
Average per graph	4.7 nodes, 4.5 edges	15.7 nodes, 16.2 edges
Maximum per graph	9 nodes, 9 edges	95 nodes, 103 edges

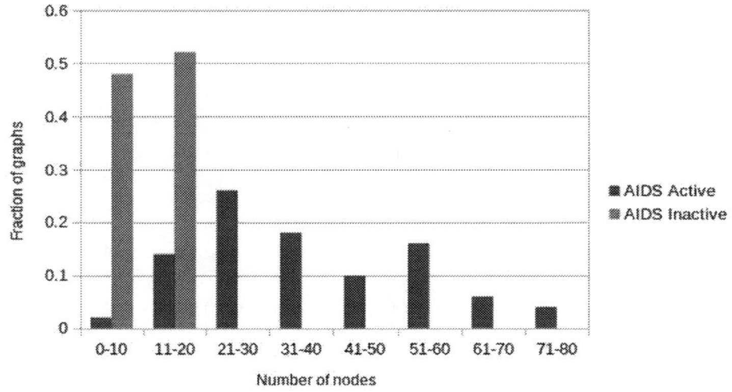

FIGURE 4.5: Fraction of graphs with their size for AIDS dataset

4.5.2.2 Execution Time Comparison

This section's results were all calculated on a system with a 3.40 GHz processor, 9.8 GB of RAM, and other hardware specifications. Figure 4.6 shows the average execution time in milliseconds for graph matching between GED and k^*-GED ($k = 1$, 2, and 3) for letter graphs using A^* with beam search optimization and a beam width of $w = 10$. Notably, running time from GED to 1^*-GED utilizing A^* is significantly shorter. The drop in 1-degree nodes in the input graphs is responsible for this execution time reduction.

Figure 4.7 shows the running time comparison between GED and k^*-GED for AIDS graphs using A^*, and Figure 4.8 shows the time in milliseconds using the beam search heuristic. The execution time in both scenarios decreases

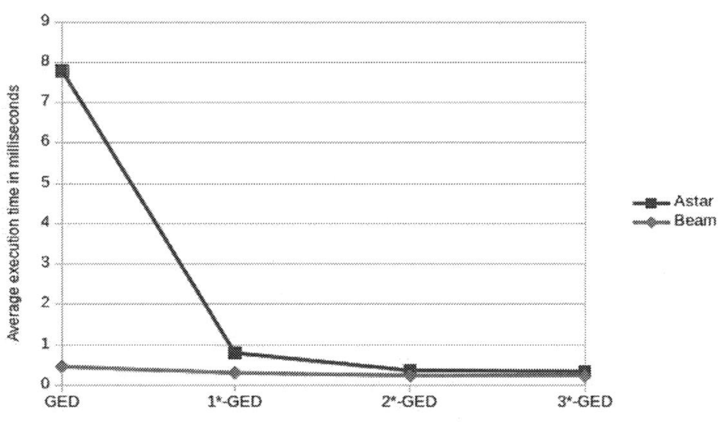

FIGURE 4.6: GED vs. k^*-GED computation on letter graphs

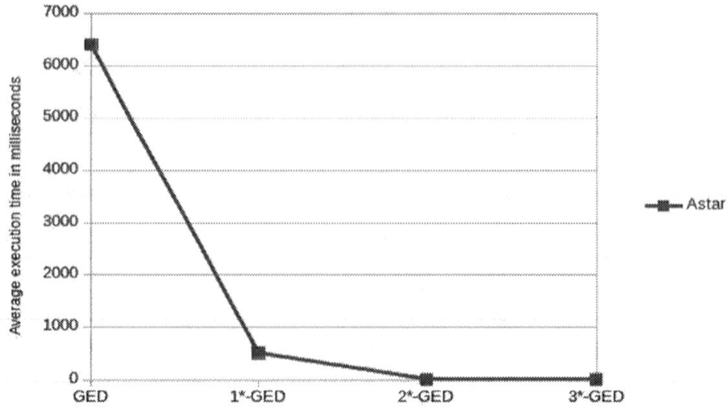

FIGURE 4.7: *GED* vs. *k*GED* computation on AIDS graphs

noticeably from GED to 1*-GED using A^*, then declines from 1*-GED to 2*-GED and then again from 2*-GED to 3*-GED. This trend is mainly explained by the fact that it is more difficult to remove 2-degree and 3-degree nodes from the network than it is to remove 1-degree nodes because doing so runs the risk of disconnecting the graph.

4.5.2.3 Accuracy Comparison

We use a nearest neighbor classifier to solve the letter graph classification problem and evaluate accuracy. With 50 graphs for each of the 15 classes in the letter dataset, there are 750 letter graphs with severe distortion in the test

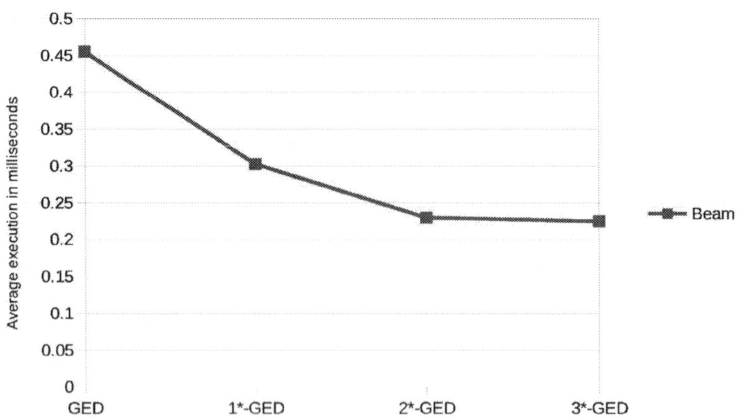

FIGURE 4.8: *GED* vs. *k*GED* computation on AIDS graphs using beam search

TABLE 4.2: Accuracy on letter dataset using GED and k^*-GED

Class	GED	1*-GED	2*-GED	3*-GED
A	94	70	64	64
E	80	54	54	54
F	80	64	44	42
H	70	42	30	30
I	94	06	06	06
K	86	44	38	38
L	86	56	36	36
M	96	90	60	60
N	86	64	52	52
T	90	50	46	44
V	90	48	46	46
W	94	78	56	54
X	78	52	50	50
Y	92	76	52	52
Z	90	80	64	64

dataset. Furthermore, there are 750 graphs in the training dataset—50 graphs for each of the 15 letters. A comparison of the classification accuracy for each of the 15 letters using k^*-GED for $k = 1$ to 3 and A^* based GED is shown in Table 4.2. In the meantime, the beam search-based classification accuracy of GED and k^*-GED is shown in Table 4.3. Here, it is seen that while using k^*-GED, letters with more straight lines typically show higher accuracy, whereas letters with fewer straight lines typically have poorer accuracy. For example, the accuracy of the letters I, L, T, and V is lower than that of the letters M, W, Y, and Z. Moreover, it is significant that k^*-GED accuracy is almost the same with or without beam search. This is explained by the fact that the edit path taken into consideration by beam search becomes equal to that of A^* since the number of nodes in the letter graphs formed after k^*-GED becomes so low.

We use a test dataset of 1500 graphs, of which 1200 graphs are from the inactive class of AIDS compounds and 300 graphs are from the active class, to evaluate accuracy on the AIDS dataset. There are 250 graphs in the training dataset: 200 graphs from the inactive class and 50 graphs from the active class. The classification accuracy for the AIDS dataset is compared in Table 4.4 between GED with the beam search heuristic and k^*-GED for $k = 1$ to 4 with the beam search heuristic, using a beam width of $w = 10$. Furthermore, Table 4.5 presents an analysis of the AIDS dataset's classification accuracy, comparing bipartite GED with bipartite k^*-GED for $k = 1$ to 4.

TABLE 4.3: Accuracy on letter dataset using GED and k^*-GED with beam

Class	GED	1^*-GED	2^*-GED	3^*-GED
A	90	70	64	64
E	86	54	54	54
F	84	64	44	42
H	60	42	30	30
I	94	06	06	06
K	80	44	38	38
L	86	56	36	36
M	94	90	60	60
N	84	64	52	52
T	88	50	46	44
V	88	48	46	46
W	92	78	56	54
X	76	52	50	50
Y	92	76	52	52
Z	90	80	64	64

TABLE 4.4: Accuracy on AIDS dataset using GED and k^*-GED with beam

Class	Active	Inactive
GED with beam	98.6	99.9
1^*-GED with beam	98.6	99.3
2^*-GED with beam	97.3	99.2
3^*-GED with beam	96.6	99.1
4^*-GED with beam	96.6	99.1

TABLE 4.5: Accuracy on AIDS dataset using Bipartite GED and k^*-GED

Class	Active	Inactive
Bipartite GED	98.6	99.9
Bipartite 1^*-GED	98.6	99.3
Bipartite 2^*-GED	97.3	99.2
Bipartite 3^*-GED	97.0	99.2
Bipartite 4^*-GED	96.6	99.1

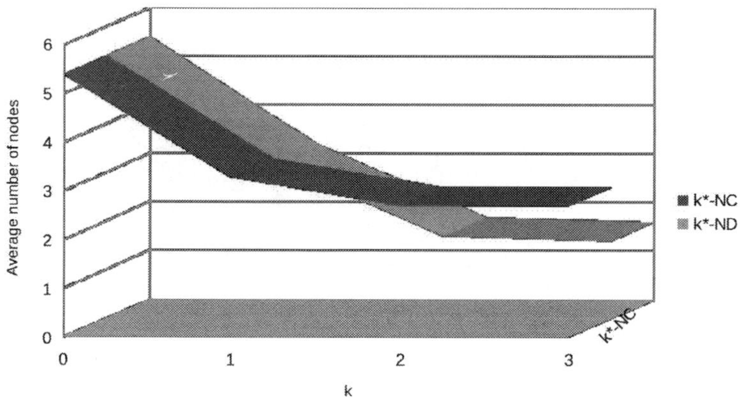

FIGURE 4.9: Effect of k^*-NC and k^*-ND on the topology of Letter A

Interestingly, the AIDS dataset's classification accuracy is higher than the letter dataset's. This increased precision highlights the usefulness of the concept for larger networks because smaller graphs typically undergo more substantial graph changes as nodes contract than larger graphs.

4.5.2.4 Topological Considerations and Discussion

The input graphs' structure is dramatically changed by the node contraction technique. Figure 4.9 shows how k^*-NC affects the training set's graph topology for the letter A. We have added the effect of k^*-ND on the structure of the same graphs for comparative study. In addition, we may investigate how k^*-NC and k^*-ND affect the graph topology in the training set of active chemicals from the AIDS dataset using Figure 4.10. Notably, k^*-ND shows a rapid decrease in average number of nodes from $k = 1$ to 3, indicating that nearly all of the graphs' structure has been deleted. Hence, since k^*-NC only slightly alters the graphs' structure, it can be viewed as a compromise between preserving efficiency and changing topology.

The suggested approach works well in scenarios when there are a large number of nodes and moderately dense network representations. One can iterate from $k = 0$ to $k = n$, depending on the trade-off between time economy and similarity precision. No matter how much time is left, $k = 0$ (the usual graph edit distance) must be used if precise matching is required. You can begin with $k = 1$ if approximate similarity is acceptable. On the other hand, if time is a crucial component, the process can start at $k = n$, where n is the lowest degree of nodes between G_1 and G_2. This technique can also be used as an online or anytime graph matching approach; it is especially helpful when a match needs to be found in a certain amount of time. It is important to remember that k^*-ND-GED, which allows nodes and associated edges to be

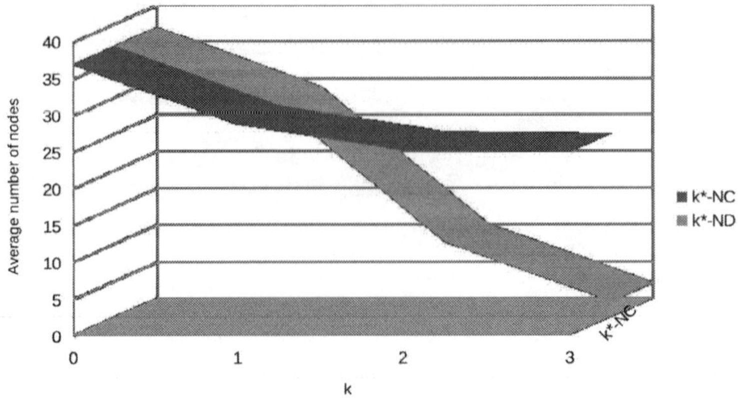

FIGURE 4.10: Effect of k^*-NC and k^*-ND on the topology of active molecules of AIDS dataset

deleted even if they are cut vertices, may provide increased efficiency at the cost of decreased accuracy.

4.6 Summary

Using the idea of graph homeomorphism, this chapter presents an error-tolerant graph matching method. We introduce the Homeomorphic-Graph-Edit-Distance algorithm, which lowers the computation of graph edit distance by reducing the number of vertices in input graphs through the use of path contraction. Furthermore, we provide a node contraction-based error-tolerant graph matching technique in which nodes are eliminated according to their degree centrality, a proxy for importance. As long as it doesn't result in an increase in the number of connected components, the removal gives priority to nodes with lower degrees. As a preparation step, the k^*-Node-Contraction is useful for several graph matching techniques. We illustrate the efficiency increases in execution time with this strategy on the AIDS and letter datasets, albeit at the expense of a modest drop in accuracy. Results on the AIDS dataset show particular promise, suggesting that the method works well for huge graphs. The chapter also explores how node contraction affects graph topology and finds that the technique becomes efficient without appreciably changing the topology of the graph as a whole.

Chapter 5

Graph Matching using Centrality Measures

5.1 Introduction

The existence of noise or error during graph processing limits the practical uses of exact graph matching in real-world circumstances. There is another method for achieving approximate graph matching: error-tolerant graph matching. Other techniques have been developed to accomplish effective graph matching at the cost of a minor drop in accuracy because of the exponential complexity linked to graph edit distance. In order to decrease the size of the matched graphs, nodes are removed according to their degree of centrality using the node contraction technique, which is covered in depth in Chapter 4. But degree centrality might not be the best standard to exclude nodes in every situation. For the purpose of removing nodes and shrinking graph sizes, choosing the right centrality measure becomes crucial depending on the characteristics and structure of various datasets.

In this chapter, in addition to degree centrality, other centrality metrics like eigenvector centrality, betweenness centrality, and PageRank centrality are examined. In order to facilitate the reduction of graph size for an early and approximative graph matching between two graphs, this multi-criteria approach seeks to customize the node removal procedure based on the datasets' features [64].

Let's review the centrality measures in brief [141]. Centrality metrics seek to pinpoint the most important or central nodes in a network. Centrality in a graph indicates the relative importance of a node inside that graph. *Degree centrality*, which reflects a node's degree, is the most basic centrality metric. When compared to nodes with fewer connections, nodes with more connections to neighboring nodes show higher degree centrality. The degree to which a node is situated on paths connecting other nodes is evaluated by *Betweenness* centrality. A generalization of degree centrality, *eigenvector* centrality gives each node a value based on the sum of the values of its neighbors. For a node u_i its eigenvector centrality is given by $x_i = \kappa_1^{-1} \sum_j A_{ij} x_j$, where κ_1 is the largest eigenvalue of adjacency matrix A and A_{ij} is an element of A. PageRank centrality ties a node's centrality to that of its neighbors, normalized by their

DOI: 10.1201/9781003532248-5

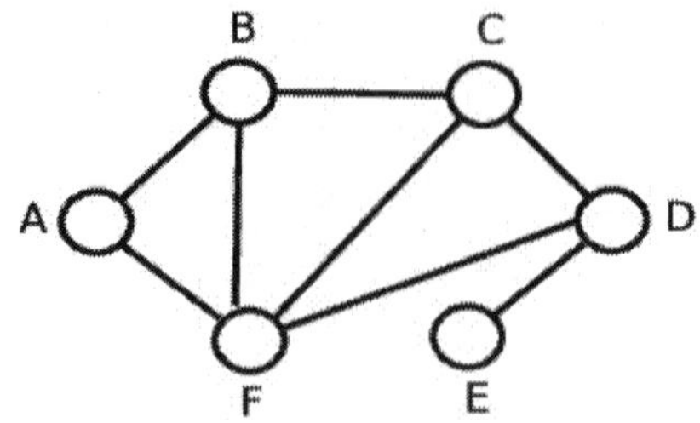

FIGURE 5.1: Graph G

outgoing degrees. The *PageRank* centrality is defined by $x_i = \alpha \sum_j A_{ij} \frac{x_j}{k_j} + \gamma$, where α is a free parameter, k_j is the outgoing degree and γ is a constant.

As an example, let's look at the graph G, which has six nodes, as shown in Figure 5.1. Nodes A, B, C, D, E, and F have degree centrality values of $2, 3, 3, 3, 1$, and 4, respectively. Given that they are not located on the shortest path between any two nodes in graph G, nodes A and E have a betweenness centrality of 0. Node B's centrality is 0.5 since it is on one of the two shortest routes connecting A and C. In a similar vein, vertices C, D, and F have betweenness centrality values of $1, 4$, and 3.5, respectively. After multiplying the degree vector by the adjacency matrix of node A, B, C, D, E, and F in graph G, the eigenvector centrality values for each node are $7, 9, 10, 8, 3$ and 11, respectively. In terms of PageRank centrality, the values for vertices A, B, C, D, E, and F are, respectively, $0.09, 0.18, 0.15, 0.26, 0.05$, and 0.25, assuming that each node of G is a web page and each edge is a bidirectional connection.

This chapter is organized as follows. Section 5.2 presents r-centrality graph edit distance and describes error-tolerant graph matching with centrality metrics. The experimental evaluations are covered in depth in Section 5.3. In conclusion, Section 5.4 provides a succinct overview.

5.2 Error-Tolerant Approximate Graph Matching using Centrality Information

In an effort to decrease the computation time for error-tolerant graph matching, we opt to exclude nodes with lower centrality values from the graphs before calculating a similarity score using Graph Edit Distance (GED) between the two graphs.

Definition 5.1 *r-centrality node contraction involves the contraction of r fractions of nodes from a graph G, specifically targeting nodes with the lowest centrality values based on a specified centrality measure.*

In this definition, r represents a ratio equivalent to the decimal form of the percentage of nodes from graph G. For instance, if $r = 0.1$, it signifies the contraction of 10% of nodes from G.

According to the definition given, a maximum of r times the total number of nodes in $G(|G|)$ are eliminated if they are not cut vertices, starting from the node in a graph G with the lowest centrality value. r-centrality node contraction (r-NC) can be expressed as r-degree centrality node contraction (r-DC-NC), r-betweenness centrality node contraction (r-BC-NC), r-eigenvector node contraction (r-EV-NC), and r-PageRank node contraction (r-PR-NC), depending on the chosen centrality measure.

Definition 5.2 *r-degree centrality node contraction involves the operation of contracting r times the total number of nodes ($|G|$) with the smallest degree from graph G.*

If r times the total number of nodes in the graph $r.|G|$ is equivalent to the number of nodes with degree k, then r-degree centrality node contraction aligns with the operation of k-degree node contraction.

Definition 5.3 *r-betweenness centrality node contraction involves the operation of contracting r times the total number of nodes ($r.|G|$) with the lowest betweenness score from graph G.*

Definition 5.4 *r-eigenvector centrality node contraction entails contracting r times the total number of nodes ($r.|G|$) with the lowest eigenvector centrality from graph G.*

Definition 5.5 *r-PageRank centrality node contraction involves contracting r times the total number of nodes ($r.|G|$) with the lowest PageRank score from graph G.*

Definition 5.6 *The computation of r-centrality GED between two graphs, G_1 and G_2, is defined as the GED between these graphs after contracting r times the total number of nodes ($r.|G1|$) from $G1$ and r times the total number of nodes ($r.|G2|$) from $G2$ with the lowest centrality values.*

The definition above characterizes the type of r-centrality Graph Edit Distance computation (r-GED) based on the particular centrality criteria that are used. These criteria correspond to different modes, including r-degree centrality GED computation (r-DC-GED), r-betweenness centrality GED computation (r-BC-GED), r-eigenvector GED computation (r-EV-GED), and r-PageRank GED computation (r-PR-GED). The usual GED computation is equivalent to r-GED when $r = 0$.

5.2.1 Edit Cost

The edit cost of r-GED can be defined by introducing an additional operation, denoted as $c(u \rightarrow \epsilon) = 0$, for r times the total number of vertices

$(r.|G|)$ in graph G with the lowest scores according to the specified centrality measure.

r-GED employs Euclidean distance, assigning a constant cost to the insertion, deletion, and substitution of vertices and edges. For two graphs G_1 and G_2, with vertices $u \in V_1$, $v \in V_2$ and edges $e \in E_1$, $f \in E_2$, the extended edit cost function is outlined as follows.

$c(u \to \epsilon) = x_{node}$
$c(\epsilon \to v) = x_{node}$
$c(u \to v) = y_{node}.||\mu_1(u) - \mu_2(v)||$
$c(e \to \epsilon) = x_{edge}$
$c(\epsilon \to f) = x_{edge}$
$c(e \to f) = y_{edge}.||\nu_1(e) - \nu_2(f)||$
$c(u \to \epsilon) = 0$, if u is one of the $r.|G|$ nodes of the lowest centrality value and is not a cut vertex.

Here x_{node}, y_{node}, x_{edge}, y_{edge} are positive constants.

5.2.2 Algorithm

The process of error-tolerant graph matching utilizing r-centrality node contraction is delineated in Algorithm 9. The algorithm takes as input two graphs, $G_1 = (V_1, E_1, \mu_1, \nu_1)$ and $G_2 = (V_2, E_2, \mu_2, \nu_2)$, along with a parameter r. Its output is the minimum cost r-GED between G_1 and G_2. In line 1, it calls Algorithm 10 to remove $\lceil r.n \rceil$ nodes in G_1 with the lowest centrality values, ensuring they are not cut vertices. Here, $\lceil\rceil$ represents the *ceiling* function, returning the smallest integer greater than or equal to rn. Similarly, line 2 removes $\lceil r.m \rceil$ nodes in G_2 with the lowest centrality values. The resulting graphs, G_1' and G_2', are obtained after performing r-Centrality-Node-Contraction on G_1 and G_2, respectively, leading to vertex sets $V_1' = \{u_1', ..., u_{n'}'\}$ and $V_2' = \{v_1', ..., v_{m'}'\}$. Line 3 initializes an empty set A. The vertex u_1' of G_1' is substituted by each vertex v_j' of G_2' in the *for* loop of lines 3–6, and the deletion of u_1' is executed in line 7. The computation of the minimum cost edit path takes place in the *while* loop of lines 8–27. The *if* loop in line 10 checks whether C_{min} is a complete edit path, ensuring the full transformation of G_1' to G_2'. If all nodes of V_1' are processed (line 13), the remaining nodes of V_2' are simply inserted in C_{min} in the *for* loop of lines 14–16. Similarly, all unprocessed vertices of V_1' are substituted by all vertices of V_2', along with the deletion of vertices of V_1' in the *for* loop of lines 19–23, and A is updated in line 24.

Algorithm 10 outlines the procedure for r-centrality node contraction. The *for* loop in lines 1–7 iterates up to $\lceil r.|G|\rceil)$ times, examining nodes in G to determine if they are cut vertices. If a node is identified as a cut vertex, it is removed along with its connected edges, and the graph G is subsequently updated.

Algorithm 9 : r-Centrality-Graph-Edit-Distance (G_1, G_2)

Require: Two Graphs G_1, G_2, where $V_1 = \{u_1, ..., u_n\}$ and $V_2 = \{v_1, ..., v_m\}$ and a parameter r
Ensure: A minimum cost r-GED between G_1 and G_2

```
G'_1 ← r-Centrality-Node-Contraction (G_1, ⌈r.n⌉)
G'_2 ← r-Centrality-Node-Contraction (G_2, ⌈r.m⌉)
A ← ∅
for each (v'_j ∈ V'_2) do
    A ← A ∪ {u'_1 → v'_j}
end for
A ← A ∪ {u'_1 → ε}
while (True) do
    Compute minimum cost edit path C_min from A
    if (C_min is a complete edit path) then
        return C_min
    else
        if (all vertices (u'_i ∈ V'_1) are visited) then
            for all unvisited (v'_j ∈ V'_2) do
                C_min ← C_min ∪ {ε → v'_j}
            end for
            A ← A ∪ {C_min}
        else
            for (all unvisited vertices (u'_i ∈ V'_1)) do
                for (each (v'_j ∈ V'_2)) do
                    C_min ← C_min ∪ {u'_i → v'_j} ∪ {u'_i → ε}
                end for
            end for
            A ← A ∪ {C_min}
        end if
    end if
end while
```

5.2.3 Computational Analysis

Proposition 5.1 *r-Centrality-Graph-Edit-Distance algorithm executes error-tolerant graph matching of G'_1 and G'_2.*

Leveraging the properties inherent in the edit costs of r-GED, Algorithm 9 yields the minimum cost of a complete edit path, facilitating the transformation of the input graph G'_1 to output graph G'_2. This ensures that each vertex of G'_1 uniquely corresponds to a vertex of G'_2. Additionally, the r-Centrality-Node-Contraction algorithm guarantees that $V'_1 \subset V_1$ and $V'_2 \subset V_2$.

Proposition 5.2 *r-Centrality-Node-Contraction algorithm operates in $O(n)$ time.*

Algorithm 10 : r-Centrality-Node-Contraction (G, r)

Require: A Graph G and a parameter r
Ensure: Transformed graph after applying r-Centrality-Node-Contraction on G

```
for (i ← 1 to ⌈r.|G|⌉) do
    Select node u with minimum centrality
    if (u is not cut vertex) then
        V ← V \ {u}
        E ← E \ {(u, v)|(u, v) ∈ E, ∀v ∈ G}
    end if
end for
return G
```

Determining whether a node u is a cut vertex can be done in $O(n)$ time. Consequently, the *for* loop in the algorithm takes $O(r.|G|.n)$ time, which simplifies to $O(n)$.

The worst-case computational complexity of the r-Centrality-Graph-Edit-Distance algorithm exhibits an exponential relationship with the number of vertices in the input graphs. To mitigate overall computation time, selecting an appropriate variable r becomes crucial.

5.2.4 t-Centrality Graph Edit Distance

Definition 5.7 *t-centrality node contraction consists of contracting t nodes from a graph G with the lowest centrality values according to a specified centrality measure.*

According to the description given above, up to t nodes in a graph G are eliminated, starting with the node with the lowest centrality value and making sure they are not cut vertices. t-centrality node contraction (t-NC) can take the form of t-degree centrality node contraction (t-DC-NC), t-betweenness centrality node contraction (t-BC-NC), t-eigenvector node contraction (t-EV-NC), or t-PageRank node contraction (t-PR-NC), depending on the centrality measure that is selected.

Definition 5.8 *The process of contracting the t nodes with the lowest degree from a graph G is implied by the term "k-degree centrality node contraction."*

If t is equal to the number of nodes with degree k in a graph, then t-degree centrality node contraction aligns with the operation of t-degree node contraction.

Definition 5.9 *The process of contracting t nodes from a graph G that have the lowest betweenness score is known as t-betweenness centrality node contraction.*

Definition 5.10 *t-eigenvector centrality node contraction entails the process of contracting t nodes with the lowest eigenvector centrality from a graph G.*

Definition 5.11 *t-PageRank centrality node contraction involves the process of contracting t nodes with the lowest PageRank score from a graph G.*

Definition 5.12 *The GED between two graphs, G_1 and G_2, taking into account the contraction of t nodes with the least centrality from both graphs, is known as the t-centrality GED computation.*

According to the aforementioned definition, t-centrality GED computation (t-GED) corresponds to a number of measures, including t-degree centrality GED computation (t-DC-GED), t-betweenness centrality GED computation (t-BC-GED), t-eigenvector GED computation (t-EV-GED), and t-PageRank GED computation (t-PR-GED), depending on the specific centrality criteria used.

Algorithm 11 outlines the computation of error-tolerant Graph Matching using t-centrality node contraction. The input to the t-Centrality-Graph-Edit-Distance algorithm includes two graphs, $G_1 = (V_1, E_1, \mu_1, \nu_1)$, $G_2 = (V_2, E_2, \mu_2, \nu_2)$, and a parameter t. The algorithm aims to determine the minimum cost t-GED between G_1 and G_2. The procedure involves invoking t-Centrality-Node-Contraction on graphs G_1 and G_2 in lines 1–2, removing t nodes with the lowest centrality values, ensuring they are not cut vertices. The resulting graphs, G_1' and G_2', are obtained after applying t-Centrality-Node-Contraction on G_1 and G_2, respectively, where $V_1' = \{u_1', ..., u_{n'}'\}$ and $V_2' = \{v_1', ..., v_{m'}'\}$. Line 3 initializes an empty set A. In the *for* loop of lines 3–6, vertex u_1' of G_1' is substituted by each vertex v_j' of G_2', and the deletion of u_1' is executed in line 7. Lines 8–27 of the *while* loop compute the minimal cost edit path. The *if* loop in line 10 verifies that G_1' is transformed into G_2' by asking if C_{min} is a complete edit path. The remaining nodes of V_2' are entered into C_{min} in the for loop of lines 14–16 if all of the nodes in V_1' have been processed (line 13). In a similar manner, in the *for* loop of lines 19–23, all unprocessed vertices of V_1' are replaced with all vertices of V_2', and in line 24, A is updated. Additionally, vertices of V_1' are deleted.

5.3 Experimental Evaluation

In this section, the r-Centrality-Graph-Edit-Distance algorithm for error-tolerant graph matching is used incorporating degree, betweenness, eigenvector, and PageRank centrality measures. Our evaluation involves utilizing the IAM graph database [148] to compare the execution time and accuracy achieved through these centrality techniques. Additionally, the proposed

Algorithm 11 : t-Centrality-Graph-Edit-Distance (G_1, G_2)

Require: Two Graphs G_1, G_2, where $V_1 = \{u_1, ..., u_n\}$ and $V_2 = \{v_1, ..., v_m\}$ and a parameter t
Ensure: A minimum cost t-GED between G_1 and G_2

$G'_1 \leftarrow t$-CENTRALITY-NODE-CONTRACTION (G_1, t)
$G'_2 \leftarrow t$-CENTRALITY-NODE-CONTRACTION (G_2, t)
$A \leftarrow \emptyset$
for each $(v'_j \in V'_2)$ **do**
  $A \leftarrow A \cup \{u'_1 \rightarrow v'_j\}$
end for
$A \leftarrow A \cup \{u'_1 \rightarrow \epsilon\}$
while (True) **do**
  Compute minimum cost edit path C_{min} from A
  if (C_{min} is a complete edit path) **then**
    return C_{min}
  else
    if (all vertices $(u'_i \in V'_1)$ are visited) **then**
      for all unvisited $(v'_j \in V'_2)$ **do**
        $C_{min} \leftarrow C_{min} \cup \{\epsilon \rightarrow v'_j\}$
      end for
      $A \leftarrow A \cup \{C_{min}\}$
    else
      for (all unvisited vertices $(u'_i \in V'_1)$) **do**
        for (each $(v'_j \in V'_2)$) **do**
          $C_{min} \leftarrow C_{min} \cup \{u'_i \rightarrow v'_j\} \cup \{u'_i \rightarrow \epsilon\}$
        end for
      end for
      $A \leftarrow A \cup \{C_{min}\}$
    end if
  end if
end while
procedure t-**Centrality-Node-Contraction**(G, t)
  for $(i \leftarrow 1$ to $t)$ **do**
    Select node u with minimum centrality
    if (u is not cut vertex) **then**
      $V \leftarrow V \setminus \{u\}$
      $E \leftarrow E \setminus \{(u, v) | (u, v) \in E, \forall v \in G\}$
    end if
  end for
  return G
end procedure

error-tolerant graph matching scheme is assessed using the Letter and AIDS datasets.

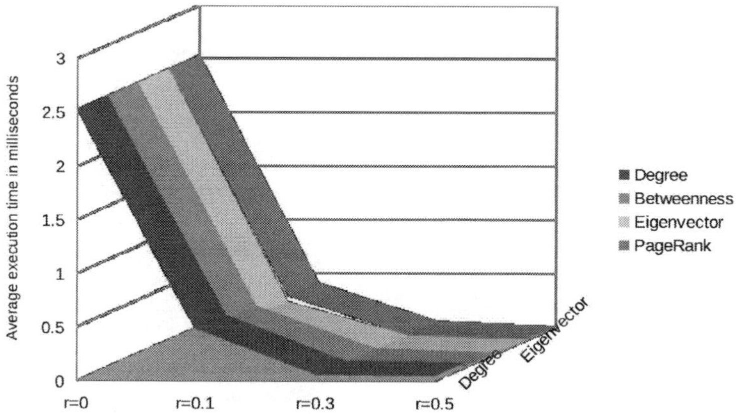

FIGURE 5.2: Comparison of execution time for letter A dataset

5.3.1 Execution Time Comparison

For comparison purposes, we employed three distinct values of r in r-GED, specifically $r = 0.1, r = 0.3$, and $r = 0.5$. Utilizing these values, we computed 0.1-GED, 0.2-GED, and 0.3-GED. The comparison of the average execution time for graph matching in milliseconds, employing the r-Centrality-Graph-Edit-Distance algorithm on the high distortion letter dataset for letters A and E, with different centrality measures, is illustrated in Figures 5.2 and Figure 5.3 respectively.

It can be observed that the graph matching time is the shortest when using the eigenvector centrality criterion, while it is comparatively longer when employing degree centrality. The computation time for the letter E is higher due to its larger number of nodes compared to the letter A.

The comparison of the average runtime for graph matching in milliseconds, employing the beam search heuristic with a beam width of $w = 10$, is presented in Figure 5.4 for the active class of the AIDS dataset, considering four different centrality measures. The figure indicates that Algorithm 9 generally requires less time when using eigenvector and betweenness centrality compared to degree and PageRank centrality.

Figure 5.5 illustrates the average execution time for graph matching on the inactive AIDS dataset, employing four different centrality measures. Once again, the computation time using eigenvector centrality is observed to be shorter than that of degree, betweenness, and PageRank centrality measures. The running time using betweenness centrality is generally lower compared to the degree and PageRank centrality measures.

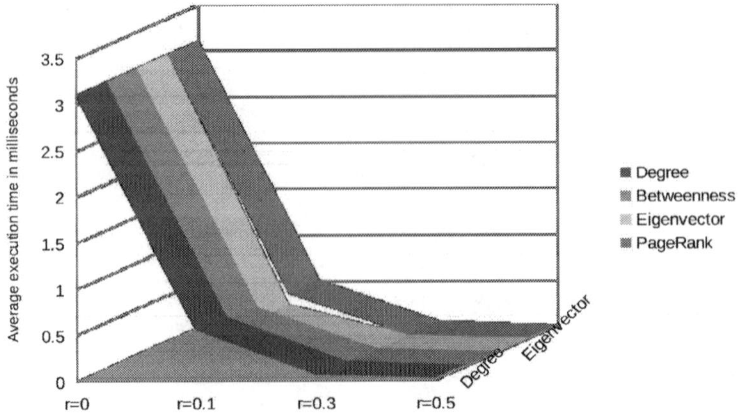

FIGURE 5.3: Comparison of execution time for letter E dataset

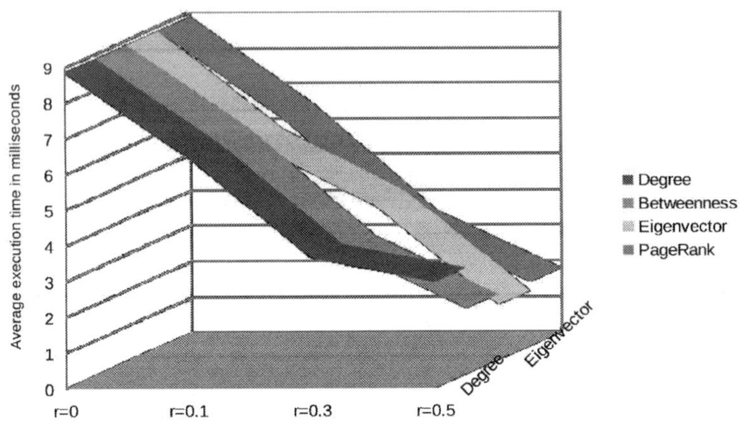

FIGURE 5.4: Comparison of execution time for active class of AIDS dataset

5.3.2 Accuracy Comparison

A nearest-neighbor classifier is used to solve the graph classification problem and assess accuracy. In the training and test sets, there are 750 graphs in the letter dataset with a high-degree of distortion. For each of the 15 letters, there are 50 graphs in each of these datasets. Using four centrality indicators, Figure 5.6 shows the classification accuracy of the suggested graph matching for letter A with high distortion. Using the same centrality criteria, Figure 5.7 shows the accuracy of graph matching for letter E.

In this case, eigenvector centrality is found to provide better accuracy for both letters A and E than the other three measures. This implies that the relative importance of nodes in the letter dataset is better captured by

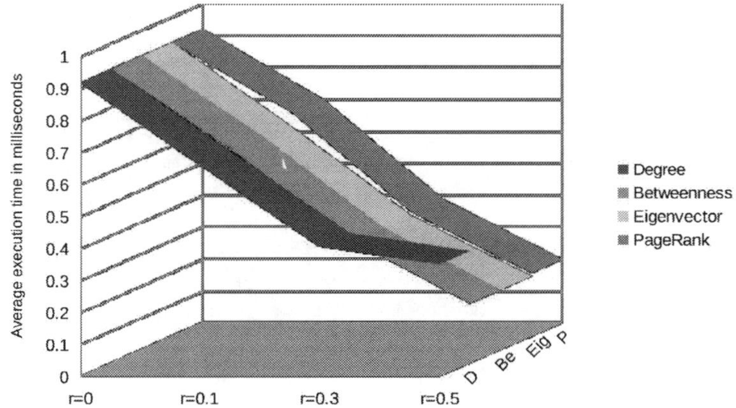

FIGURE 5.5: Comparison of execution time for inactive class of AIDS dataset

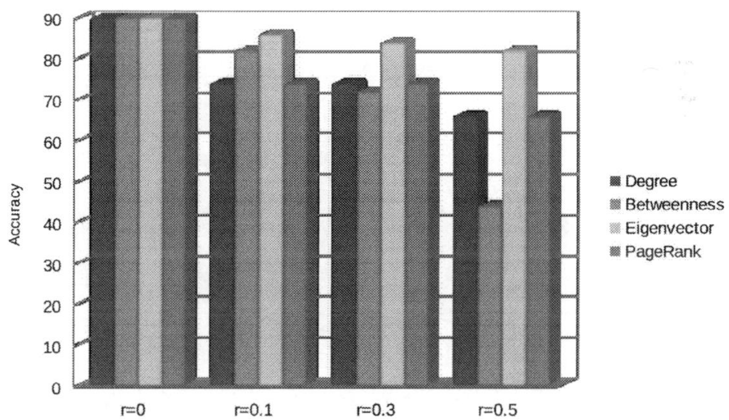

FIGURE 5.6: Comparison of accuracy ratio of letter A dataset

eigenvector centrality. The accuracy scores for each of the fifteen letters using degree centrality for r values of $0.1, 0.3$, and 0.5 are shown in Table 5.1 for more comprehensive accuracy findings. Furthermore, accuracy results for all letters based on betweenness, eigenvector, and PageRank centrality are shown in Tables 5.2, 5.3, and 5.4, respectively.

A test dataset of 1200 graphs from the inactive class and 300 graphs from the active class are used to assess accuracy on the AIDS dataset. The training dataset comprises 200 graphs from the AIDS dataset's inactive class and 50 graphs from the dataset's active class. Figure 5.8 shows the accuracy ratio of the suggested error-tolerant strategy for the active class of the AIDS dataset using four different centrality measures. This figure shows that when

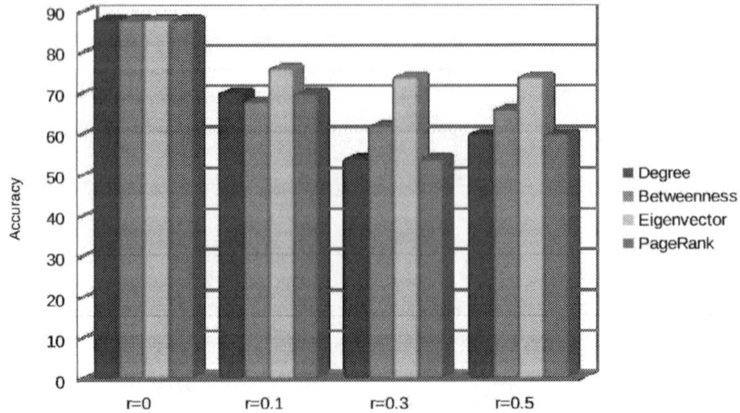

FIGURE 5.7: Comparison of accuracy ratio of letter E dataset

TABLE 5.1: Accuracy on letter dataset using degree centrality

Class	$r = 0.1$	$r = 0.2$	$r = 0.3$
A	74	74	66
E	70	54	60
F	72	44	42
H	58	42	26
I	90	86	84
K	68	58	44
L	78	64	54
M	82	50	66
N	58	50	50
T	70	46	52
V	88	54	60
W	88	74	60
X	76	52	40
Y	80	62	52
Z	78	62	64

employing betweenness centrality, the accuracy ratio for the AIDS dataset's active class is typically greater than when using alternative centrality measures.

Figure 5.9 depicts the comparison of accuracy for the inactive class of the AIDS dataset using the four centrality measures. In this figure, both the degree and PageRank criteria demonstrate higher accuracy for the classification of graphs in the AIDS dataset.

TABLE 5.2: Accuracy on letter dataset using betweenness centrality

Class	$r = 0.1$	$r = 0.2$	$r = 0.3$
A	82	72	44
E	68	62	66
F	62	54	40
H	54	38	40
I	94	92	74
K	76	36	38
L	82	68	52
M	86	70	32
N	62	50	30
T	68	56	62
V	78	52	80
W	84	60	46
X	58	48	34
Y	82	76	74
Z	62	50	48

TABLE 5.3: Accuracy on letter dataset using eigenvector centrality

Class	$r = 0.1$	$r = 0.2$	$r = 0.3$
A	86	84	82
E	99	96	92
F	58	42	22
H	40	32	42
I	96	92	58
K	68	38	42
L	76	60	56
M	84	54	30
N	64	38	30
T	54	34	36
V	72	36	66
W	90	62	34
X	62	34	34
Y	74	80	72
Z	62	56	36

TABLE 5.4: Accuracy on letter dataset using PageRank centrality

Class	$r = 0.1$	$r = 0.2$	$r = 0.3$
A	74	74	66
E	70	54	60
F	72	44	42
H	58	42	26
I	90	86	84
K	68	58	44
L	78	64	54
M	82	50	66
N	58	50	50
T	70	46	52
V	88	54	60
W	88	74	60
X	76	52	40
Y	80	62	52
Z	78	62	64

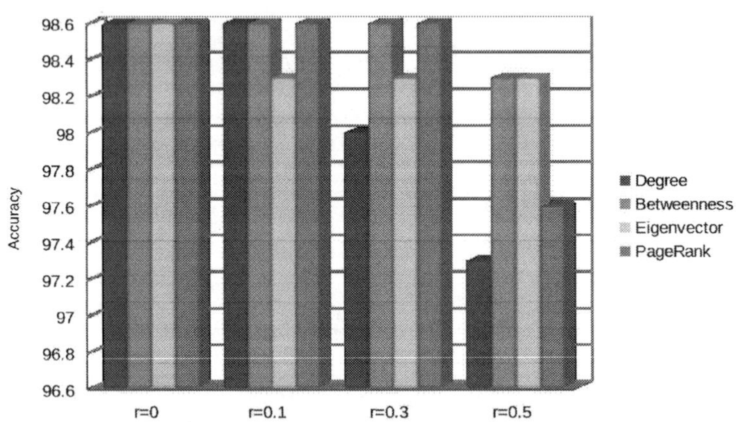

FIGURE 5.8: Comparison of accuracy for active class of AIDS dataset

Tables 5.5–5.8 present the accuracy ratio of the AIDS dataset using degree, betweenness, eigenvector, and PageRank centrality for three distinct values of $r = 0.1, 0.3$, and 0.5.

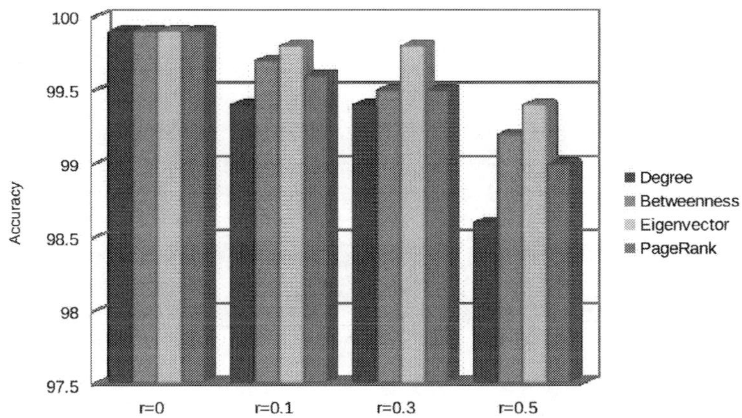

FIGURE 5.9: Comparison of accuracy for inactive class of AIDS dataset

TABLE 5.5: Accuracy on AIDS dataset using degree centrality

Class	$r = 0.1$	$r = 0.2$	$r = 0.3$
Active	98.6	98	97.3
Inactive	99.4	99.4	98.6

TABLE 5.6: Accuracy on AIDS dataset using betweenness centrality

Class	$r = 0.1$	$r = 0.2$	$r = 0.3$
Active	98.6	98.6	98.3
Inactive	99.7	99.5	99.2

TABLE 5.7: Accuracy on AIDS dataset using eigenvector centrality

Class	$r = 0.1$	$r = 0.2$	$r = 0.3$
Active	98.3	98.3	98.3
Inactive	99.8	99.8	99.4

5.3.3 t-Centrality Graph Edit Distance

The number of nodes taken into consideration for contraction in k^*-degree node contraction is the same as the value of t^* in t^*-GED, for comparison's

TABLE 5.8: Accuracy on AIDS dataset using PageRank centrality

Class	$r = 0.1$	$r = 0.2$	$r = 0.3$
Active	98.6	98.6	97.6
Inactive	99.6	99.5	99

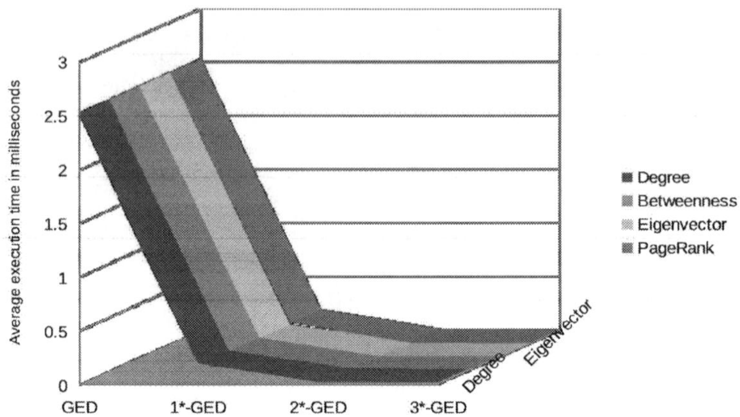

FIGURE 5.10: Comparison of execution time for letter A dataset

sake. Thus, 1^* in 1^*-GED indicates the number of nodes with degree 1, 2^* in 2^*-GED indicates the number of nodes with degrees 1 and 2, and 3^* in 3^*-GED indicates the number of nodes with degrees 1, 2, and 3. Figures 5.10 and 5.11, respectively, show the average execution time of the t-Centrality-Graph-Edit-Distance algorithm for graph matching for letters A and E of the high-distortion letter dataset when different centrality measures are applied.

Notice that the graph matching time is minimized when employing eigenvector criteria, while the time using degree centrality is comparatively higher. The computation time for letter E is elevated due to its larger number of nodes compared to letter A.

Figure 5.12 illustrates the comparison of the average runtime for graph matching in milliseconds using beam search heuristic (with a beam width $w = 10$) across four different centrality measures for the active class of the AIDS dataset. It is evident from this figure that Algorithm 11 generally requires less time when utilizing eigenvector and betweenness centrality, as opposed to degree and PageRank centrality.

Figure 5.13 depicts the average execution time for graph matching on the inactive AIDS dataset using four centrality measures. Once again, the computation time is notably lower when employing eigenvector and betweenness

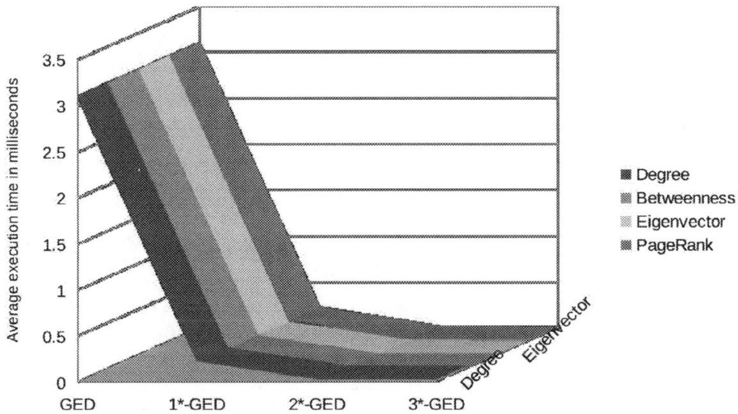

FIGURE 5.11: Comparison of execution time for letter E dataset

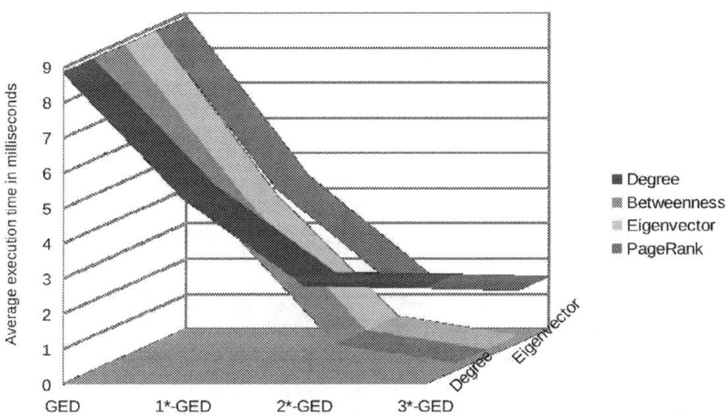

FIGURE 5.12: Comparison of execution time for active class of AIDS dataset

criteria compared to degree and PageRank. Additionally, between the latter two, the average time using PageRank is less than that with degree centrality.

For accuracy evaluation, we address the task of classifying graphs using the nearest neighbor classifier. The letter dataset with a high-distortion level comprises 750 graphs for both training and test sets. Each of these datasets contains 50 graphs for each of the 15 letters. Figure 5.14, illustrates the classification accuracy of the proposed graph matching for letter A with high distortion using four centrality indicators, while Figure 5.15, displays the accuracy of graph matching for letter E with the same measures.

In this context, it is noteworthy that the accuracy of letter A, when using degree centrality, is comparatively lower than the other three measures

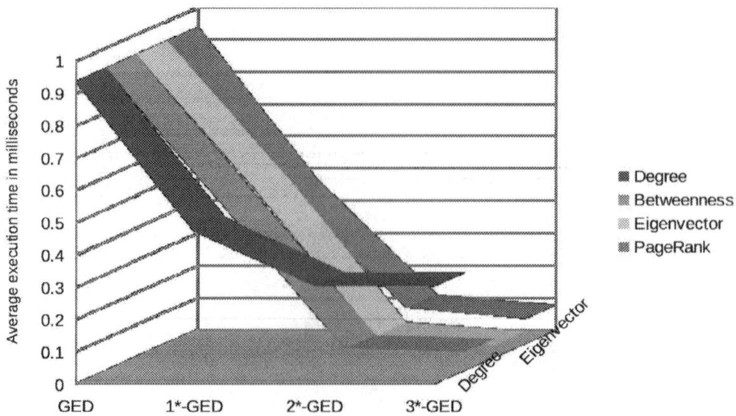

FIGURE 5.13: Comparison of execution time for inactive class of AIDS dataset

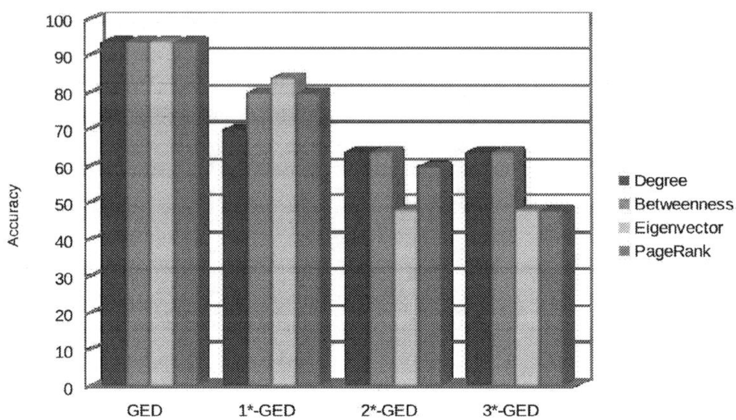

FIGURE 5.14: Comparison of accuracy ratio of letter A dataset

when contracting t nodes. Here, t corresponds to nodes with degree 1 in the input graphs (1^*-GED). Additionally, the accuracy ratio for letter E tends to be higher when employing betweenness and PageRank, even though these centrality measures exhibit lower computation time than degree centrality.

To assess accuracy on the AIDS dataset, we employed a test dataset consisting of 300 graphs from the active class and 1200 graphs from the inactive class. The training dataset comprises 50 graphs from the active class and 200 graphs from the inactive class of the AIDS dataset. Figure 5.16 illustrates the accuracy ratio of the proposed error-tolerant scheme using four different centrality measures. Notably, in this figure, we observe that accuracy obtained with degree and PageRank centrality is generally higher than that achieved

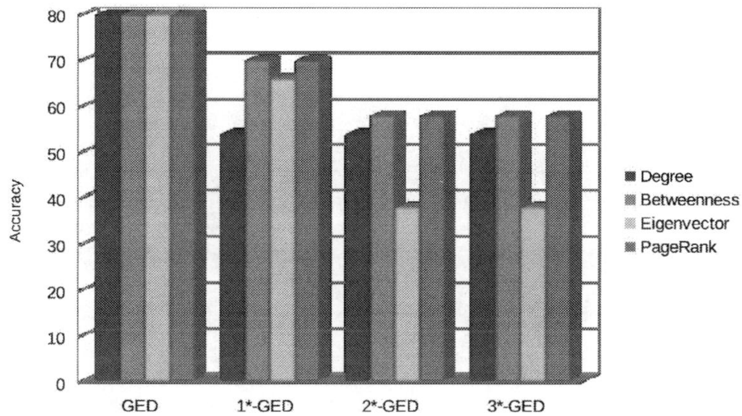

FIGURE 5.15: Comparison of accuracy ratio of letter E dataset

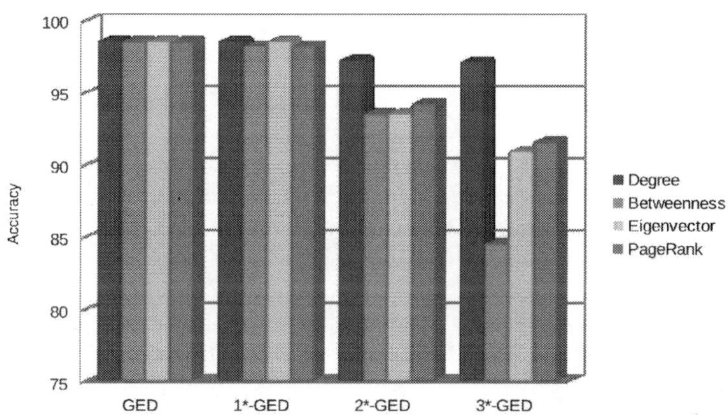

FIGURE 5.16: Comparison of accuracy for active class of AIDS dataset

with eigenvector and betweenness centrality. This highlights a trade-off between time and accuracy; centrality criteria with lower computation time tend to result in less accuracy, while more accurate centrality techniques require additional computation time.

Figure 5.17 presents a comparison of accuracy for the inactive class of the AIDS dataset using four centrality measures. Once again, both degree and PageRank criteria demonstrate higher accuracy in the classification of graphs within the AIDS dataset.

Based on the aforementioned experiments, it is evident that node contraction using different centrality measures can yield varying impacts on both execution time and accuracy ratios within a specific graph dataset. Consequently, the selection of the most appropriate centrality measure for a given

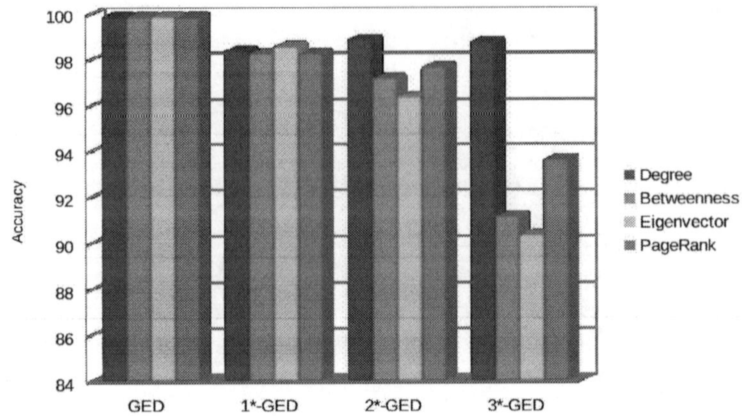

FIGURE 5.17: Comparison of accuracy for inactive class of AIDS dataset

dataset can be tailored to meet specific requirements, balancing considerations related to execution time and accuracy.

5.4 Summary

In this chapter, methodologies are introduced aimed at approximating graph matching through the utilization of centrality measures. The core idea involves diminishing the graph sizes by disregarding nodes with lower values based on specified centrality criteria. Alongside degree centrality, we explored the application of eigenvector, betweenness, and PageRank centrality measures to facilitate node contraction in the context of error-tolerant graph matching computations.

The experimental findings highlighted the versatile nature of these centrality criteria, portraying them as valuable tools for striking a balance between computation time and accuracy in diverse graph datasets. Through systematic evaluations, we observed that eigenvector, betweenness, and PageRank centrality, in addition to degree centrality, can be leveraged as effective trade-offs between computation efficiency and matching precision in the realm of error-tolerant graph matching.

Chapter 6

Geometric Graph Matching

6.1 Introduction

Geometric graph matching is the process of evaluating the similarity between geometric graphs, where every vertex has an associated coordinate point in a plane. Exact matching computes a strict correspondence between two geometric graphs, whereas in error-tolerant matching an approximate similarity between two geometric graphs is calculated. In this chapter, we present an approach to error-tolerant graph matching using geometric graphs. We introduce the vertex distance or dissimilarity and edge distance or dissimilarity between two geometric graphs and use it to compute graph distance. Finally, we use graph distance to perform exact and error-tolerant geometric graph matching.

A graph is said to be a geometric graph where each of the vertices is assigned a coordinate point in a two-dimensional or three-dimensional plane. Geometric graph matching is the process of computing the similarity between the two geometric graphs. Geometric graph matching is a particular type of graph matching in which each input graph is modeled as a geometric graph. Depending on the nature of the matching, the graph matching process is broadly divided into exact and error-tolerant graph matching. Exact graph matching requires a strict correspondence between nodes and edges of the two geometric graphs. It is like a geometric graph isomorphism, which in addition to preserving the structure of the two geometric graphs, also preserves the coordinates of the corresponding vertices.

Exact geometric graph matching, although theoretically appealing, may not be useful in many real-world applications, as due to the presence of noise or distortion during the processing, the input graph data may be altered. In such situations, we use error-tolerant geometric graph matching, which is also known as inexact graph matching due to its flexibility to accommodate errors during the process of matching [37].

As each vertex of a geometric graph has a unique coordinate representation, we can use this information along with other geometric transformation properties to perform geometric graph matching. Kuramochi and Karypis [114] in 2007 proposed an algorithm to compute geometric graph isomorphism in polynomial time. The 2009 paper by Cheong *et al.* [32] have shown geometric graph matching using edit distance method to be NP-hard.

DOI: 10.1201/9781003532248-6

Approximate solution for geometric graph matching using vertex edit distance is described in [6]. In [145] authors have described geometric graph matching by applying Monte Carlo tree search.

The graph matching problem seems to be inherently linked to the geometry and topology of graphs. This chapter describes a simple yet promising framework for geometric graph similarity and matching [61]. A graph matching algorithm is introduced using geometric graphs, which is both error-tolerant as well as computable in polynomial time [63]. The vertex distance is defined between two geometric graphs (not between two vertices) using the position of their vertices, and it is shown to be metric over the set of all graphs with vertices only. Then edge distance is defined between two graphs based on the angular orientation and length of the edges. A modified version of edge distance is also defined, which uses the position of edges as an additional attribute, and demonstrate it to be a metric. Then the notion of vertex distance and edge distance are combined to define the graph distance between two geometric graphs and show some of its properties. Finally, this graph distance is used to perform error-tolerant graph matching on letter and AIDS dataset, in which all the nodes have a coordinate position in the two-dimensional plane [63].

This chapter is organized as follows. Section 6.2 provides mathematical foundation. Section 6.3 presents the novel geometric graph similarity framework. Section 6.4, introduce geometric graph matching based on the proposed geometric graph similarity framework. It describes both exact as well as error-tolerant graph matching. Section 6.5 contains an experimental evaluation of the proposed geometric graph matching scheme. Finally, section 6.6 provides a summary.

6.2 Mathematical Foundation

In this section, we review the basic definitions and notations used in this chapter. We also explain the motivation for the described work.

A *geometric graph* G is defined as $G = (V, E, l, c)$, where V is the set of nodes, E is the set of edges, l is a labeling function $l : \{V \cup E\} \rightarrow \Sigma$ which assigns a label from Σ to each vertex and edge, c is a function $c : V \rightarrow \mathbb{R}^2$ which assigns a coordinate point to each vertex of G. If $\Sigma = \emptyset$ then G is called the unlabeled geometric graph.

Let X be a nonempty set. A mapping $d : X \times X \rightarrow \mathbb{R}$ is defined to be a *metric* on X if d satisfy the following conditions:

(i) $d(u, v) \geq 0$ for all $u, v \in X$ and $d(u, v) = 0$ if and only if $u = v$

(ii) $d(u, v) = d(v, u)$ for all $u, v \in X$ (Symmetry)

(iii) $d(u, w) \leq d(u, v) + d(v, w)$ for all $u, v, w \in X$ (Triangle inequality). The pair (X, d) is called a *metric space.*

In this chapter, we consider the problem of finding a similarity score between two geometric graphs. Given two geometric graphs G_1,G_2 in graph domain $\mathcal{G}$ of the two-dimensional space, we would like to find a distance $d : \mathcal{G} \times \mathcal{G} \rightarrow \mathbb{R}$ which computes the similarity between G_1 and G_2. For general graphs, it is also known as graph comparison problem. In this chapter, we consider graph comparison problem synonymous to graph matching problem.

Edit-distance-based approach to graph matching, as well as geometric graph matching, have no efficient exact solution. For example, graph edit distance [200] and geometric graph matching based on edit distances [32] are NP-hard and therefore no efficient algorithms are available. Since our goal is not to reconstruct G_2 from G_1, therefore we can avoid the intractability of edit distance based methods for the problem of geometric graph similarity. Instead, we use the Linear Sum Assignment Problem (LSAP) formulation-based approach to geometric graph distance using the various attributes of vertices and edges of the geometric graphs.

Geometric graph isomorphism, as opposed to graph isomorphism, can be solved efficiently in polynomial time [114] so we can take advantage of this to find the best geometric representation of a geometric graph G_2 with respect to an input geometric graph G_1 which make them similar. We define geometric graph similarity which intuitively captures the minimum error required to make the two geometric graphs similar.

Initially, two geometric graphs G_1 and G_2 are considered with vertices only without connections. To find similarity between these graphs, one approach is to assign each vertex of G_1 to one of the vertices of G_2, so that the total distance is minimum. Suppose vertex coordinate points of G_1 be $\{(a_1, b_1), (a_2, b_2), \ldots, (a_n, b_n)\}$ and coordinate points of G_2 be $\{(x_1, y_1), (x_2, y_2), \ldots, (x_n, y_n)\}$, as shown in Figure 6.1. If $l_i =$

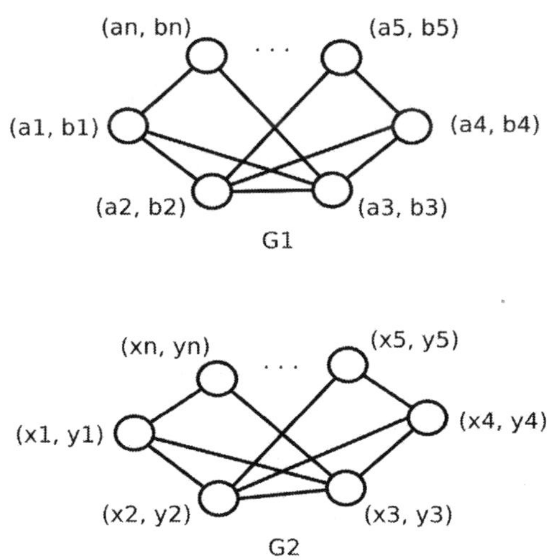

FIGURE 6.1: Geometric Graphs G_1 and G_2

$\{\sqrt{(a_i - x_j)^2 + (b_i - y_j)^2} | 1 \leq j \leq n\}, \forall 1 \leq i \leq n$. Then vertex distance between G_1 and G_2 is $\sum_{i=1}^{n} l_i = \sum_{i=1}^{n} \min_{1 \leq j \leq n}\{\sqrt{(a_i - x_j)^2 + (b_i - y_j)^2}\}$. One limitation of this expression is that more than one vertices of G_1 can be assigned to the same vertex of G_2.

For a more realistic optimization problem formulation, let $C = (c_{ij})$ be a $n \times n$ cost matrix, where c_{ij} is the Euclidean distance from i-th vertex of G_1 to j-th vertex of G_2. We have, $c_{ij} = \sqrt{(a_i - x_j)^2 + (b_i - y_j)^2}$. Now, we can minimize the objective function $\sum_{i=1}^{n} c_{i\varphi(i)}$ to get the minimum distance between G_1 and G_2 as $\min_{\varphi \in S_n} \sum_{i=1}^{n} c_{i\varphi(i)}$. Where S_n is the set of all assignments φ from V_1 of G_1 to V_2 of G_2. This is the standard LSAP formulation of vertex assignments from V_1 to V_2. The LSAP can be solved using Hungarian algorithm in $O(n^3)$ time [27].

6.3 Geometric Graph Similarity

In this section, vertex distance, edge distance between the geometric graphs G_1 and G_2 are introduced. These distance measures are utilized to compute the dissimilarity between two geometric graphs.

In the following definition of vertex distance, the number of vertices in two graphs are assumed to be equal. For arbitrary graph in which the number of vertices is not same, additional vertices can be inserted in the smaller graph, with coordinates equal to mean of all its existing coordinate values, to make the vertex set of both graphs equal.

Definition 6.1 *Let $G_1 = (V_1, E_1, c_1)$ and $G_2 = (V_2, E_2, c_2)$ be two geometric graphs with $|V_1| = |V_2| = n$. Let coordinate points of V_1 be $\{(a_1, b_1), (a_2, b_2), \ldots, (a_n, b_n)\}$ and coordinate points of V_2 be $\{(x_1, y_1), (x_2, y_2), \ldots, (x_n, y_n)\}$ then the vertex distance between the two geometric graphs G_1 and G_2 is defined as*

$$VD(G_1, G_2) = \min_{\varphi \in S_n} \sum_{i=1}^{n} \sqrt{(a_i - x_{\varphi(i)})^2 + (b_i - y_{\varphi(i)})^2} \tag{6.1}$$

Where S_n and φ are as defined above.

Here, *VD* represents an LSAP formulation for the assignment of the vertex set V_1 of G_1 to the vertex set V_2 of G_2. The geometric graphs having a similar position of vertices will have less *VD*, whereas the graphs having a very different position of vertices have large *VD*. In the above definition, we can also represent vertex set V_1 and V_2 using their offsets from the mean position of all vertex coordinates of G_1 and G_2, respectively.

Lemma 6.1 *$VD(G_1, G_2)$ is a metric on the set of all geometric graphs $\mathcal{G}$ having edge set $E = \emptyset$.*

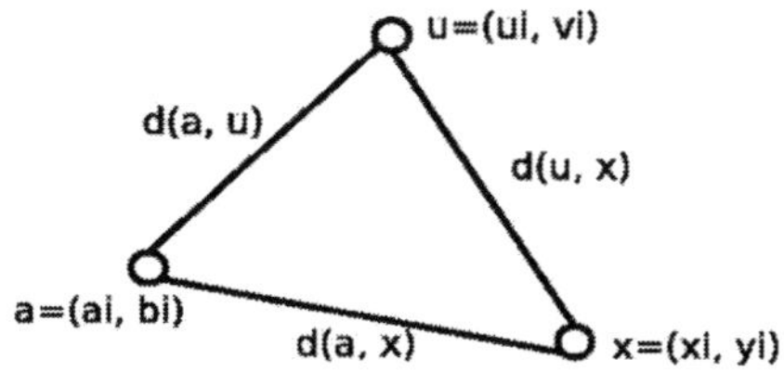

FIGURE 6.2: Triangle inequality: $d(a,x) \leq d(a,u) + d(u,x)$

Proof *Property (i) of metric: Here by definition* $VD(G_1,G_2) \geq 0$*, If* $G_1 = G_2$ *then clearly* $VD(G_1,G_2) = \min_{\varphi\in S_n}\sum_{i=1}^{n}[(a_i - x_{\varphi(i)})^2 + (b_i - y_{\varphi(i)})^2]^{1/2} = 0$*, since in each component of summation we have* $a_i = x_{\varphi(i)}$ *and* $b_i = y_{\varphi(i)}$*.*
Conversely, if $VD(G_1,G_2) = 0$ *then* $\min_{\varphi\in S_n}\sum_{i=1}^{n}[(a_i - x_{\varphi(i)})^2 + (b_i - y_{\varphi(i)})^2]^{1/2} = 0$ *implies that for some combination of* $i,\varphi(i)$ *each term of summation*
$\sum_{i=1}^{n}[(a_i - x_{\varphi(i)})^2 + (b_i - y_{\varphi(i)})^2]^{1/2}$ *is* 0,
$\Rightarrow (a_i - x_{\varphi(i)})^2 = 0$ *and* $(b_i - y_{\varphi(i)})^2 = 0$,
$\Rightarrow a_i = x_{\varphi(i)}$ *and* $b_i = y_{\varphi(i)}$*. Hence* $G_1 = G_2$
Property (ii):
$VD(G_1,G_2) = \min_{\varphi\in S_n}\sum_{i=1}^{n}[(a_i - x_{\varphi(i)})^2 + (b_i - y_{\varphi(i)})^2]^{1/2}$
$= \min_{\varphi\in S_n}\sum_{i=1}^{n}[(x_{\varphi(i)} - a_i)^2 + (y_{\varphi(i)} - b_i)^2]^{1/2} = VD(G_2,G_1)$
Reversibility of φ *can also be demonstrated. Let* $V_1 = u_1,u_2,\cdots,u_n$ *and* $V_1 = v_1,v_2,\cdots,v_n$*. Let* ϕ *from* $V_1 \to V_2$ *using LSAP assign* $u_1 \to v_1, u_2 \to v_2,\cdots,u_n \to v_n$*. Now suppose* ϕ^{-1} *from* $V_2 \to V_1$ *assigns* $v_1 \to x_1, v_2 \to x_2,\cdots,v_n \to x_n$*. There are three cases to consider.*
Case 1: Cost of ϕ *from* $V_1 \to V_2$ *= cost of* ϕ^{-1} *from* $V_2 \to V_1$ *then* $VD(G_1,G_2) = VD(G_2,G_1)$*.*
Case 2: Cost of ϕ *from* $V_1 \to V_2$ *>* *cost of* ϕ^{-1} *from* $V_2 \to V_1$ *Then* $x_1 \neq u_1, x_2 \neq u_2,\cdots,x_n \neq u_n$*. It is a contradiction since cost using* ϕ *was supposed to be minimum using LSAP. Here the cost is a distance. So* ϕ *could have selected* x_1 *instead of* u_1*,* x_2 *instead of* u_2 *and so on, to return a lower value.*
Case 3: Cost of ϕ *from* $V_1 \to V_2$ *<* *cost of* ϕ^{-1} *from* $V_2 \to V_1$*. Arguments similar to Case 2 applies here.*
Property (iii):
Let $G_3 = (V_3,E_3,c_3)$ *be a geometric graph with* $|V_3| = n$*,* $E_3 = \emptyset$ *and coordinate points of* V_3 *be* $\{(u_1,v_1),(u_2,v_2),\ldots,(u_n,v_n)\}$*. Figure 6.2 illustrate geometrically the triangle inequality among the points* $(a_i,b_i),(x_i,y_i)$ *and* (u_i,v_i) *of the graphs* G_1,G_2 *and* G_3 *respectively. We can observe that*
$VD(G_1,G_3) + VD(G_3,G_2)$
$= \min_{\varphi\in S_n}\sum_{i=1}^{n}[(a_i - u_{\varphi(i)})^2 + (b_i - v_{\varphi(i)})^2]^{1/2}$
$\quad + \min_{\varphi'\in S_n}\sum_{i=1}^{n}[(u_i - x_{\varphi'(i)})^2 + (v_i - y_{\varphi'(i)})^2]^{1/2}$
let the minimum function over all $\varphi(i)$ *and* $\varphi'(i)$ *returns the following*
$= \sum_{i=1}^{n}[(a_i - u_{\varphi(i)})^2 + (b_i - v_{\varphi(i)})^2]^{1/2} + \sum_{i=1}^{n}[(u_i - x_{\varphi'(i)})^2 + (v_i - y_{\varphi'(i)})^2]^{1/2}$

In the first term of the above expression, for $i = 1$, *let* $\varphi(1) = k$. *In the second term suppose* $u_{\varphi(1)} = u_k$ *in* G_3 *is assigned to* $x_{\varphi'(k)}$ *for some* $x \in G_2$. *So, for* $i = 1$ *we have the above expression*
$[(a_1 - u_{\varphi(1)})^2 + (b_1 - v_{\varphi(1)})^2]^{1/2} + [(u_k - x_{\varphi'(k)})^2 + (v_k - y_{\varphi'(k)})^2]^{1/2}$
$= [(a_1 - u_k)^2 + (b_1 - v_k)^2]^{1/2} + [(u_k - x_{\varphi'(k)})^2 + (v_k - y_{\varphi'(k)})^2]^{1/2}$
let $p_1 = a_1 - u_k, p_2 = b_1 - v_k,$
$q_1 = u_k - x_{\varphi'(k)}, q_2 = v_k - y_{\varphi'(k)}$, *then we have*
$[(a_1 - u_k)^2 + (b_1 - v_k)^2]^{1/2} + [(u_k - x_{\varphi'(k)})^2 + (v_k - y_{\varphi'(k)})^2]^{1/2}$
$= [(p_1)^2 + (p_2)^2]^{1/2} + [(q_1)^2 + (q_2)^2]^{1/2}$
using Minkowski inequality
$\sum_{i=1}^{n}[(x_i + y_i)^2]^{1/2} \leq (\sum_{i=1}^{n} x_i^2)^{1/2} + (\sum_{i=1}^{n} y_i^2)^{1/2}$
and putting the values of p_1, q_1, p_2, q_2 *we get*
$[(p_1)^2 + (p_2)^2]^{1/2} + [(q_1)^2 + (q_2)^2]^{1/2}$
$\geq= [(p_1 + q_1)^2 + (p_2 + q_2)^2]^{1/2}$
$= [(a_1 - u_k + u_k - x_{\varphi'(k)})^2 + (b_1 - v_k + v_k - y_{\varphi'(k)})^2]^{1/2}$
$= [(a_1 - x_{\varphi'(k)})^2 + (b_1 - y_{\varphi'(k)})^2]^{1/2}$
taking sum over all i-th term and a $\varphi'' \in S_n$
$\sum_{i=1}^{n}[(a_i - x_{\varphi''(i)})^2 + (b_i - y_{\varphi''(i)})^2]^{1/2}$
and taking minimum of summation over all i *and a* $\varphi''(i)$ *we get*
$\min_{\varphi'' \in S_n} \sum_{i=1}^{n}[(a_i - x_{\varphi''(i)})^2 + (b_i - y_{\varphi''(i)})^2]^{1/2} = VD(G_1, G_2)$
hence, $VD(G_1, G_3) + VD(G_3, G_2) \geq VD(G_1, G_2)$
Note that if either $G_3 = G_1$ *or* $G_3 = G_2$ *then*
$VD(G_1, G_3) + VD(G_3, G_2) = VD(G_1, G_2)$
and when $G_3 \neq G_1$ *and* $G_3 \neq G_2$ *then*
$VD(G_1, G_3) + VD(G_3, G_2) > VD(G_1, G_2)$.

VD considers only the position of vertices of geometric graphs in the plane. Now, we use the features of edges of geometric graphs to define the edge distance. The main features of an edge of a geometric graph include its orientation and length along with the coordinates of its endpoints.

Let $\theta_{\{(a,b),(c,d)\}}$ denote the angle subtended between the extended line joining the coordinate points $(a, b), (c, d)$ and the positive x-axis. In the following definition of edge distance, we assume that the number of edges in the two graphs is equal. For graphs with unequal edge set, we can append additional empty edges to the smaller graph, to make the total number of edges in both graphs same.

Definition 6.2 *Let* $G_1 = (V_1, E_1, c_1)$ *and* $G_2 = (V_2, E_2, c_2)$ *be two geometric graphs with* $|E_1| = |E_2| = m$. *Here* c_1 *and* c_2 *are the set of coordinate points of vertex set* V_1 *and* V_2 *respectively. Then the edge distance between the two geometric graphs* G_1 *and* G_2 *is defined as*

$$ED(G_1, G_2) = \min_{\varphi \in S_m} \sum_{i=1}^{m} c_{i\varphi(i)} \tag{6.2}$$

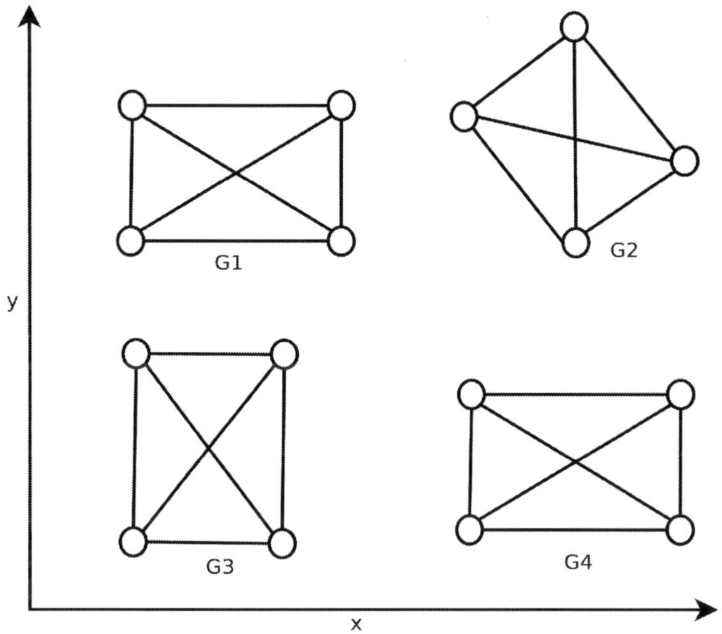

FIGURE 6.3: Edge Distance from G_1 to G_4 is 0

such that,

$$c_{ij} = \sqrt{((\Theta_i - \Theta_j)\frac{\pi}{180^\circ})^2} + \sqrt{(d_i - D_j)^2}$$

Where, $\Theta_i = \theta_{\{(a_i,b_i),(a_p,b_p)|((a_i,b_i),(a_p,b_p))\in E_1\}}$,
$\Theta_j = \theta_{\{(x_j,y_j),(x_q,y_q)|((x_j,y_j),(x_q,y_q))\in E_2\}}$,
$d_i = \{\sqrt{(a_i - a_p)^2 + (b_i - b_p)^2} | ((a_i, b_i), (a_p, b_p)) \in E_1\}$,
$D_j = \{\sqrt{(x_j - x_q)^2 + (y_j - y_q)^2} | ((x_j, y_j), (x_q, y_q)) \in E_2\}$
and S_m *is the set of all assignments* φ *from* E_1 *of* G_1 *to* E_2 *of* G_2.

The above definition of ED finds the best assignment of edges from E_1 to E_2. The cost function $C = (c_{ij})$ is itself made up of two components. The first term $\sqrt{((\Theta_i - \Theta_j)\frac{\pi}{180^\circ})^2}$ calculates difference in orientation of E_i in E_1 and E'_j in E_2. We denote this term of ED by E^A_{ij}, where A stands for an angular difference. It is translation and scaling invariant, as we can observe in Figure 6.3 that G_1 and G_4 have identical value of this term. The second term $\sqrt{(d_i - D_j)^2}$ of ED denotes the difference of length between E_i and E'_j. We call this component as the length term and represent it by E^L_{ij}, where L is used for the length of edges. E^L_{ij} is translation and rotation invariant, as we can see in Figure 6.3 that all four graphs G_1 to G_4 have the equivalent value of the second term of ED. Following properties of ED are easy to establish.

Lemma 6.2 *$ED(G_1, G_2)$ satisfy the following properties:*
(i) $ED(G_1, G_2) \geq 0$, if $G_1 = G_2$ then $ED(G_1, G_2) = 0$
(ii) $ED(G_1, G_2) = ED(G_2, G_1)$
(iii) $ED(G_1, G_2) \leq ED(G_1, G_3) + ED(G_3, G_2)$

Proof *(i) $ED(G_1, G_2) \geq 0$, since both terms in the definition of ED is always non-negative. Also when $G_1 = G_2$, we have $\Theta_i = \Theta_j$ and $d_i = D_j$ for each term of summation. Therefore $ED(G_1, G_2) = 0$. Note that converse may not be true i.e., $ED(G_1, G_2) = 0$ not necessarily imply $G_1 = G_2$. We can observe in Figure 6.3 that graphs G_2 and G_3 are rotation followed by translation of G_1, while G_4 is a translation of $G1$. Here $ED(G_1, G_4) = 0$. Thus ED between two translated version of a graph turns out to be 0.*
(ii) $ED(G_1, G_2) = \min_{\varphi \in S_m} \sum_{i=1}^{m}([((\Theta_i - \Theta_{\varphi(i)})\frac{\pi}{180°})^2]^{1/2} + [(d_i - D_{\varphi(i)})^2]^{1/2}) = \min_{\varphi \in S_m} \sum_{i=1}^{m}([((\Theta_{\varphi(i)} - \Theta_i)\frac{\pi}{180°})^2]^{1/2} + [(D_{\varphi(i)} - d_i)^2]^{1/2}) = ED(G_2, G_1)$
(iii) Let $G_3 = (V_3, E_3, c_3)$ be a geometric graph with $|V_3| = n$ and coordinate points of V_3 be $\{(u_1, v_1), (u_2, v_2), \ldots, (u_n, v_n)\}$. Suppose $\Theta_k = \theta_{\{(u_k,v_k),(u_r,v_r)|(u_k,v_k),(u_r,v_r) \in E_3\}}$ and $D_k = \{\sqrt{(u_k - u_r)^2 + (v_k - v_r)^2}|(u_k, v_k), (u_r, v_r) \in E_3\}$, then $ED(G_1, G_3) + ED(G_3, G_2)$
$= \min_{\varphi \in S_m} \sum_{i=1}^{m}([((\Theta_i - \Theta_k)\frac{\pi}{180°})^2]^{1/2} + [(d_i - D_k)^2]^{1/2})$
$\quad + \min_{\varphi \in S_m} \sum_{k=1}^{m}([((\Theta_k - \Theta_j)\frac{\pi}{180°})^2]^{1/2} + [(D_k - D_j)^2]^{1/2})$
$= \min_{\varphi \in S_m} \sum_{i=1}^{m}(|(\Theta_i - \Theta_k)\frac{\pi}{180°}| + |d_i - D_k|)$
$\quad + \min_{\varphi \in S_m} \sum_{k=1}^{m}(|(\Theta_k - \Theta_j)\frac{\pi}{180°}| + |D_k - D_j|)$
Let the minimum function over all i and $\varphi(i)$ returns the following
$\sum_{i=1}^{m}(|(\Theta_i - \Theta_k)\frac{\pi}{180°}| + |d_i - D_k|) + \sum_{k=1}^{m}(|(\Theta_k - \Theta_j)\frac{\pi}{180°}| + |D_k - D_j|)$
now, using triangle inequality $|x| + |y| \geq |x + y|$
and inequality of summation $\sum_{x \in X} |f(x)| \geq |\sum_{x \in X} f(x)|$
$\sum_{i=1}^{m}(|(\Theta_i - \Theta_k)\frac{\pi}{180°}| + |d_i - D_k|) + \sum_{k=1}^{m}(|(\Theta_k - \Theta_j)\frac{\pi}{180°}| + |D_k - D_j|)$
$= \sum_{i=1}^{m} |(\Theta_i - \Theta_k)\frac{\pi}{180°}| + \sum_{i=1}^{m} |d_i - D_k| + \sum_{k=1}^{m} |(\Theta_k - \Theta_j)\frac{\pi}{180°}| + \sum_{k=1}^{m} |D_k - D_j|$
$\geq \sum_{i,k=1}^{m}(|(\Theta_i - \Theta_k + \Theta_k - \Theta_j)\frac{\pi}{180°}| + |d_i - D_k + D_k - D_j|)$
$= \sum_{i=1}^{m}(|(\Theta_i - \Theta_j)\frac{\pi}{180°}| + |d_i - D_j|)$
and therefore $ED(G_1, G_3) + ED(G_3, G_2) \geq ED(G_1, G_2)$.

Observe that VD express the relative position of the vertices between two geometric graphs, whereas ED captures the orientation and length of edges between two geometric graphs. Nevertheless, ED does not encapsulate the relative position of edges of graphs in the two-dimensional plane. For example in Figure 6.4, two geometric graphs G_1 and G_2 are shown such that $VD(G_1, G_2) = ED(G_1, G_2) = 0$ even though geometrically $G_1 \neq G_2$, because ED does not capture the relative position of edges. To accommodate the relative position of edges, we introduce a third term in the definition of ED called position component denoted by E_{ij}^P.

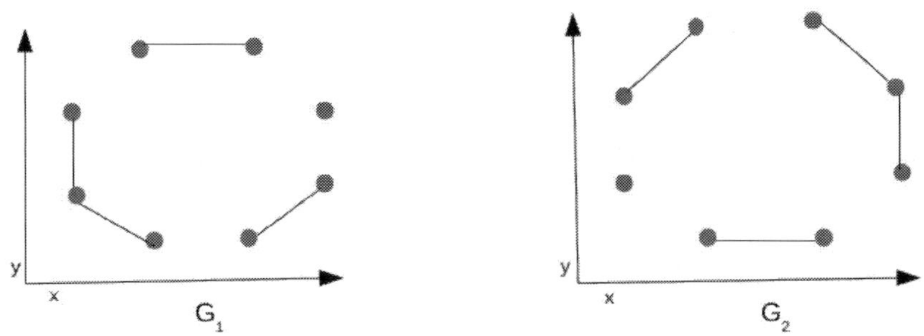

FIGURE 6.4: Geometric graphs with VD and ED both 0

Let i and j denote the i-th and j-th edge of G_1 and G_2 respectively, such that i-th edge of G_1 is $\{(a_i, b_i), (a_p, b_p)\}$ and j-th edge of G_2 is $\{(x_j, y_j), (x_q, y_q)\}$. We define the position term of ED by

$$E_{ij}^P = (\sqrt{(a_i - x_j)^2 + (b_i - y_j)}^2 + \sqrt{(a_p - x_q)^2 + (b_p - y_q)}^2)/2$$

The term E_{ij}^P is similar to VD, which associate the position of one vertex of G_1 to another vertex of G_2. E_{ij}^P compares position of one edge of G_1 to another edge of G_2. Now, we define Edge Distance Modified or Edge Distance Metric (EDM), which use an objective function to combine all three features of edges.

Definition 6.3 *Let $G_1 = (V_1, E_1, c_1)$ and $G_2 = (V_2, E_2, c_2)$ be two geometric graphs with $|E_1| = |E_2| = m$. Then the edge distance metric between the two geometric graphs G_1 and G_2 is defined as*

$$EDM(G_1, G_2) = \min_{\varphi \in S_m} \sum_{i=1}^{m} c_{i\varphi(i)} \tag{6.3}$$

such that,

$$c_{ij} = E_{ij}^A + E_{ij}^L + E_{ij}^P$$

Where, $E_{ij}^A = \sqrt{((\Theta_i - \Theta_j)\frac{\pi}{180^\circ})^2}$, $E_{ij}^L = \sqrt{(d_i - D_j)^2}$,
$E_{ij}^P = (\sqrt{(a_i - x_j)^2 + (b_i - y_j)}^2 + \sqrt{(a_p - x_q)^2 + (b_p - y_q)}^2)/2$
and S_m is the set of all assignments φ from E_1 of G_1 to E_2 of G_2.

Lemma 6.3 *$EDM(G_1, G_2)$ is metric over the set of all geometric graphs $\mathcal{G}$ without isolated vertices.*

Proof *Property (i) of metric: $EDM(G_1, G_2) \geq 0$, since E_{ij}^A, E_{ij}^L, and E_{ij}^P are all non-negative. Now, when $G_1 = G_2$, we have $c_{ij} = E_{ij}^A + E_{ij}^L + E_{ij}^P = 0$,*

hence $EDM(G_1, G_2) = 0$. *Conversely, suppose* $EDM(G_1, G_2) = 0$. *Then for some* $\varphi \in S_m$, *we get* $\sum_{i=1}^{m} c_{i\varphi(i)} = 0$. *So,* $E_{ij}^A = E_{ij}^L = E_{ij}^P = 0$. *We notice that* $E_{ij}^A = 0$ *implies* $\Theta_i = \Theta_j^{'}$ *and since* E_{ij}^A *is translation and scaling invariant therefore* $E_{ij}^A = 0$ *indicates that rotation is preserved between* G_1 *and* G_2. *Also* E_{ij}^L *is translation and rotation invariant so* $E_{ij}^L = 0$ *denotes that scaling is preserved. Similarly,* $E_{ij}^P = 0$ *implies* $a_i = x_j, b_i = y_j, a_p = x_q, b_p = y_q$ *and therefore translation is also preserved between* G_1 *and* G_2. *Hence* $EDM(G_1, G_2) = 0$ *implies that* $G_1 = G_2$ *as long as* G_1 *and* G_2 *does not have 0-degree vertices. For example, if the position of the isolated vertex of graph* G_1 *in Figure 6.4 is changed to construct another graph* G_1' *then also we get* $EDM(G_1, G_1') = 0$.
Property (ii): In the definition of EDM, *we have* $c_{ij} = E_{ij}^A + E_{ij}^L + E_{ij}^P = E_{ji}^A + E_{ji}^L + E_{ji}^P = c_{ji}$. *Hence* $EDM(G_1, G_2) = EDM(G_2, G_1)$.
Property (iii): Let $G_3 = (V_3, E_3, c_3)$ *be a geometric graph as defined in Lemma 2 having* $|E_3| = m$, *then*
$EDM(G_1, G_3) + EDM(G_3, G_2) = \min_{\varphi \in S_m} \sum_{i=1}^{m} ([((\Theta_i - \Theta_k)\frac{\pi}{180^\circ})^2]^{1/2}$
$+[(d_i - D_k)^2]^{1/2} + (\sqrt{(a_i - u_k)^2 + (b_i - v_k)}^2 + \sqrt{(a_p - u_r)^2 + (b_p - v_r)}^2)/2)$
$+ \min_{\varphi \in S_m} \sum_{k=1}^{m} ([((\Theta_k - \Theta_j)\frac{\pi}{180^\circ})^2]^{1/2} + [(D_k - D_j)^2]^{1/2}$
$+(\sqrt{(u_k - x_j)^2 + (v_k - y_j)}^2 + \sqrt{(u_r - x_q)^2 + (v_r - y_q)}^2)/2)$
$= \min_{\varphi \in S_m} \sum_{i=1}^{m} (E_{ik}^A + E_{ik}^L + E_{ik}^P) + \min_{\varphi \in S_m} (\sum_{k=1}^{m} E_{kj}^A + E_{kj}^L + E_{kj}^P)$ *using Lemma 2 and the fact that each* E_{ij}^P *is itself a Euclidean distance function, we have*
$\min_{\varphi \in S_m} \sum_{i,k=1}^{m} [(E_{ik}^A + E_{kj}^A) + (E_{ik}^L + E_{kj}^L) + (E_{ik}^P + E_{kj}^P)] \geq \sum_{i=1}^{m} (E_{ij}^A + E_{ij}^L + E_{ij}^P) = EDM(G_1, G_2)$

For the purpose of showing the triangle inequality property of VD, ED or EDM; we can use the properties of their cost matrix. We can observe that $EDM(G_1, G_2)$ is the minimum cost of assignment of edges from G_1 to G_2 using its cost function. Similarly $EDM(G_1, G_3)$ and $EDM(G_3, G_2)$ are the minimum cost of selecting edges from G_1 to G_3 and G_3 to G_2 respectively. So, $EDM(G_1, G_3) + EDM(G_3, G_2)$ is a valid assignment from G_1 to G_2. Since, $EDM(G_1, G_2)$ pertains to assignment of edges from G_1 to G_2 with minimum cost, we have $EDM(G_1, G_2) \leq EDM(G_1, G_3) + EDM(G_3, G_2)$.

Graph distance between two geometric graphs is defined as a linear combination of both VD and ED for the same pair of graphs.

Definition 6.4 *Let* $G_1 = (V_1, E_1, c_1)$ *and* $G_2 = (V_2, E_2, c_2)$ *be two geometric graphs with* $|V_1| = |V_2| = n$ *and* $|E_1| = |E_2| = m$. *Then the graph distance between the two geometric graphs* G_1 *and* G_2 *is defined as*

$$GD(G_1, G_2) = VD(G_1, G_2) + ED(G_1, G_2). \tag{6.4}$$

GD combines both VD and ED. We call the first component in the definition of GD as VD-term and the second component as ED-term.

Theorem 1 *$GD(G_1, G_2)$ satisfy the following properties:*
(i) $GD(G_1, G_2) \geq 0$, if $G_1 = G_2$ then $GD(G_1, G_2) = 0$
(ii) $GD(G_1, G_2) = GD(G_2, G_1)$
(iii) $GD(G_1, G_2) \leq GD(G_1, G_3) + GD(G_3, G_2)$

The above follows from Lemma 6.1 and Lemma 6.2. Since both VD-term and ED-term separately satisfy the above properties.

Definition 6.5 *Let $G_1 = (V_1, E_1, c_1)$ and $G_2 = (V_2, E_2, c_2)$ be two geometric graphs with $|V_1| = |V_2| = n$ and $|E_1| = |E_2| = m$. Then the Graph Distance Modified or Graph Distance Metric (GDM) between the two geometric graphs G_1 and G_2 is defined as*

$$GDM(G_1, G_2) = VD(G_1, G_2) + EDM(G_1, G_2). \tag{6.5}$$

In the above definition of graph distance modified we use EDM instead of ED for the computation of edge distance between two graphs.

Theorem 2 *$GDM(G_1, G_2)$ is a metric over set of all geometric graphs $\mathcal{G}$.*

Proof *From Lemma 6.1 and Lemma 6.3, we observe that both VD-term and EDM-term of GDM satisfy properties (ii), (iii) of metric. Now, we consider property (i):*
Here certainly, $GDM(G_1, G_2) \geq 0$ since $VD(G_1, G_2) \geq 0$ and $EDM(G_1, G_2) \geq 0$.
If $G_1 = G_2$ then $VD(G_1, G_2) = 0$ and $EDM(G_1, G_2) = 0$ and therefore $GDM(G_1, G_2) = 0$.
Now if $GDM(G_1, G_2) = 0$ then both VD-term and EDM-term are 0. When VD-term is 0, i.e., $VD(G_1, G_2) = 0$, then from Lemma 1, $G_1 = G_2$ without considering the links, in other words positions of all matched vertices between G_1 and G_2 coincide. When EDM-term is 0, then from Lemma 6.3, $E^A_{ij} = E^L_{ij} = E^P_{ij} = 0$ and therefore, angular orientation, length and positions of each matched edges are exactly the same. By combining the above two facts we get $G_1 = G_2$, whenever $GDM(G_1, G_2) = 0$.

6.4 Geometric Graph Matching

6.4.1 Exact Geometric Graph Matching

In this section, we describe an algorithm to check whether two geometric graphs are isomorphic. Before checking the isomorphism between two geometric graphs, their reference coordinate must be identical. Therefore, first, we carry out graph alignment of two input graphs so that their reference coordinates become identical. Steps to perform graph alignment of G_2 with respect to G_1 is described in Algorithm 12. The input to this algorithm is two geometric graphs G_1 and G_2 and output is the geometric transform of G_2 with the maximum number of edges having similar orientation to G_1. Line 1 of this algorithm selects an edge e of E_1 to be a reference axis. The *for* loop in line 2 of the algorithm choose an edge f of E_2 to perform geometric-transform on G_2. The function geometric-transform process the coordinates of G_2 using a transformation matrix to make edge f to have same orientation to reference axis e with the left coordinate of edge f coincident to left coordinate of e. This particular geometric configuration of G_2 will be translation and rotation invariant to G_1. To make this geometric configuration scaling invariant also, we make the edges e and f of unit length and uniformly scale the other edges of G_1 and G_2. Line 4 of the algorithm computes ED between G_1 and G'_2 and in line 6, the algorithm updates d_{min} to find the minimum ED over all the geometric configuration of G_2. Line 10 of the algorithm returns the graph G corresponding to minimum ED value.

Algorithm 12 :Graph-Alignment(G_1, G_2)

Require: Two undirected geometric graphs G_1, G_2, where $G_1 = (V_1, E_1, c_1), G_2 = (V_2, E_2, c_2)$ with $|V_1| = |V_2| = n, |E_1| = |E_2| = |m|$
Ensure: Geometric transform of G_2 with maximum edges aligned to G_1

1: Select an edge $e \in E_1$ as a reference axis
2: **for** (each edge $f \in E_2$) **do**
3: $\quad G'_2 \leftarrow$ geometric-transform(G_2)
4: $\quad d \leftarrow ED(G_1, G'_2)$
5: $\quad$ **if** $(d < d_{min})$ **then**
6: $\quad\quad d_{min} \leftarrow d$
7: $\quad\quad G \leftarrow G'_2$
8: $\quad$ **end if**
9: **end for**
10: **return** (G)

Test to perform geometric graph isomorphism between two graphs is presented in Algorithm 13. The input to this algorithm is two geometric graphs G_1 and G_2. Output to this algorithm is isomorphic graphs when G_1 is isomorphic to G_2; t-tolerant isomorphic graphs when G_1 is t-tolerant isomorphic to G_2; and graph distance GD when G_1 is neither isomorphic to G_2 nor t-tolerant isomorphic to G_2. Line 1 of Geometric-Graph-Isomorphism algorithm calls Algorithm 12 to perform graph alignment of G_2 with respect to G_1. Vertex assignments from G_1 to G_2 using the definition of VD is given in line 2. We can use the Hungarian algorithm for optimal assignment of vertices. Based on the assignment VD is calculated in line 3. Angle and length component of ED is initialized in lines 4–5. Line 7 computes ED based on edge assignments from E_1 to E_2. Final GD value is calculated in line 8. The *if* loop in line 9 checks whether GD is 0, and line 10 also check uniqueness of edge assignments

Algorithm 13 :Geometric-Graph-Isomorphism(G_1, G_2)

Require: Two undirected geometric graphs G_1, G_2, where $G_1 = (V_1, E_1, c_1), G_2 = (V_2, E_2, c_2)$ with $|V_1| = |V_2| = n, |E_1| = |E_2| = |m|$

Ensure: Isomorphic graphs or t-tolerant isomorphic graphs or graph distance between G_1 and G_2

1: $G_2 \leftarrow$ Graph-Alignment(G_1, G_2)
2: Compute vertex assignment from V_1 to V_2
3: $VD \leftarrow \sum_{i=1}^{n} \sqrt{(a_i - x_{\varphi(i)})^2 + (b_i - y_{\varphi(i)})^2}$
4: $E_{ij}^{A} \leftarrow \sqrt{((\Theta_{ij} - \Theta'_{ij})\frac{\pi}{180^\circ})^2}$
5: $E_{ij}^{L} \leftarrow \sqrt{(d_{ij} - D_{ij})^2}$
6: Compute edge assignment from E_1 to E_2
7: $ED \leftarrow \sum_{i=1}^{n}(E_{i\varphi(i)}^{A} + E_{i\varphi(i)}^{L})$
8: $GD \leftarrow VD + ED$
9: **if** $(GD == 0)$
10: Check uniqueness of edge assignments **then**
11: **return** G_1 is isomorphic to G_2
12: **else if**
13: **for** each $i = 1$ to n **do**
14: $(a_i - x_i) < t$ **and** $(b_i - y_i) < t$
15: **end for**
16: Check uniqueness of edge assignments **then**
17: **return** G_1 is t-tolerant isomorphic to G_2
18: **else**
19: **return** (GD)
20: **end if**

which ensures that each assigned edge from G_1 to G_2 connects the same set of coordinate points there by implying a valid geometric graph isomorphism. The algorithm then returns in line 11, indicating that G_1 is isomorphic to G_2. The *for* loop in the lines 13–15 of algorithm checks if all the assigned coordinates from G_1 to G_2 are within distance t, and again it checks uniqueness of edge assignments, it then returns in line 17 showing that G_1 is t-tolerant isomorphic to G_2. When GD is neither 0 nor the assigned coordinates from G_1 to G_2 are within specified distance t, the algorithm outputs the GD value in line 19.

Proposition 6.1 *Graph-Alignment algorithm executes in* $O(m.n^3)$ *time.*

The geometric-transform operation in line 3 of Algorithm 12 takes $O(n^2)$ time, once a transformation matrix is constructed. The ED operation in line 4 is performed in $O(n^3)$ time, hence total time taken by this algorithm is $O(m.n^3)$.

Proposition 6.2 *Geometric-Graph-Isomorphism algorithm executes in* $O(n^3 + m^3)$ *time exclusive of Algorithm 12 in line 1.*

The assignment of vertices in line 2 can be performed in $O(n^3)$ using Munkres or Hungarian algorithm. Similarly, the assignment of edges in line 6 can be achieved in $O(n^3)$. Remaining steps can be computed in $O(n^2)$ time. Therefore overall execution time becomes $O(n^3 + m^3)$ time.

6.4.2 Error-Tolerant Geometric Graph Matching

The computation of graph distance between two geometric graphs G_1 and G_2 is described in Algorithm 14. The input to the algorithm is two undirected geometric graphs G_1 and G_2 and four weighting parameters w_1, w_2, w_3, and w_4, which are application dependent. G_1 has n_1 vertices and m_1 edges, whereas G_2 has n_2 vertices and m_2 edges. For simplicity, we have assumed that $n_1 \geq n_2$. The output of the algorithm is the geometric graph distance between G_1 and G_2.

Algorithm 14 :Geometric-Graph-Distance(G_1, G_2)

Require: Two undirected geometric graphs G_1, G_2, where $G_1 = (V_1, E_1, c_1), |V_1| = n_1, |E_1| = m_1, G_2 = (V_2, E_2, c_2), |V_2| = n_2, |E_2| = |m_2|$ and weighting factors w_i for $i = 1$ to 4; assume $n_1 \geq n_2$

Ensure: Geometric graph distance between G_1 and G_2

$(x, y) \leftarrow ((\sum_{i=1}^{n_2} x_i)/n_2, (\sum_{i=1}^{n_2} y_i)/n_2)$
if $(n_1 > n_2)$ **then**
    for $i = 0$ to $(n_1 - n_2)$ **do**
        $V_2 \leftarrow V_2 \cup \{(x, y)\}$
    end for
end if
if $(m_1 > m_2)$ **then**
    Append $(m_1 - m_2)$ empty edges to E_2
else
    Append $(m_2 - m_1)$ empty edges to E_1
end if
$G_2 \leftarrow$ Graph-Alignment-Modified(G_1, G_2) ▷ preprocessing step
Compute vertex assignment φ from V_1 to V_2
$VD \leftarrow \sum_{i=1}^{n} \sqrt{(a_i - x_{\varphi(i)})^2 + (b_i - y_{\varphi(i)})^2}$
Compute edge assignment φ from E_1 to E_2
$E_{ij}^A \leftarrow \sqrt{((\Theta_i - \Theta_j)\frac{\pi}{180^\circ})^2}$; $E_{ij}^L \leftarrow \sqrt{(d_i - D_j)^2}$
$E_{ij}^P \leftarrow \sqrt{(a_i - x_j)^2 + (b_i - y_j)}^2 + \sqrt{(a_p - x_q)^2 + (b_p - y_q)}^2$
$E_{ij}^P \leftarrow (E_{ij}^P)/2$
$EDM \leftarrow \sum_{i=1}^{n} (w_2 \cdot E_{i\varphi(i)}^A + w_3 \cdot E_{i\varphi(i)}^L + w_4 \cdot E_{i\varphi(i)}^P)$
$GDM \leftarrow w_1 \cdot VD + EDM$
return (GDM)

When the number of vertices in $|V_1| = n_1$ is greater than that of $|V_2| = n_2$, then line 1 of the Geometric-Graph-Distance algorithm computes the mean values of the x and y coordinates of graph G_2 as $x = (\sum_{i=1}^{n_2} x_i)/n_2$, $y = (\sum_{i=1}^{n_2} y_i)/n_2$ and $n_1 - n_2$ vertices with these coordinate positions are appended to V_2 in line 4 of the *for* loop in lines 3–5 to make $|V_1| = |V_2| = n$. Similarly, *if* loop in lines 7–11 check the number of edges in G_1 and G_2 and it makes them equal by appending empty edges with length and angle value both 0 to E_2 or E_1 depending on whether $m_1 > m_2$ or $m_1 < m_2$, respectively.

One optional step of this algorithm in line 12 is the preprocessing phase for graph alignment to make their reference coordinates identical. Geometric-Graph-Distance algorithm computes vertex assignment from V_1 to V_2 based on the cost function given in the definition of VD in line 13. We can use Munkres or Hungarian algorithm for optimal assignment of vertices. Based on the assignment VD is calculated in line 14. Similarly, line 15 computes edge assignment from E_1 to E_2 using the cost function given in the definition of EDM. Line 19 computes EDM by multiplying weighting factor w_2 to orientation term E_{ij}^A, w_3 to length term E_{ij}^L and w_4 to position term E_{ij}^P. Finally, line 20 computes GDM by multiplying w_1 to VD and adding it to EDM.

Steps to perform graph alignment of G_2 with respect to G_1 is described in Graph-Alignment-Modified algorithm. This algorithm is same as Graph-Alignment algorithm except that it calls $EDM(G_1, G_2')$ instead of $ED(G_1, G_2')$ in line 4 to find the minimum value of d.

6.4.3 Computational Analysis

Proposition 6.3 *Geometric-Graph-Distance algorithm executes in $O(n^3 + m^3)$ time.*

The assignment of vertices in line 13 can be performed in $O(max\{n_1, n_2\}^3)$ using Munkres or Hungarian algorithm. Similarly, the assignment of edges in line 15 can be achieved in $O(max\{m_1, m_2\}^3)$. Remaining steps can be computed in $O(n^2)$ time. Let $n = max(n_1, n_2)$ and $m = max(m_1, m_2)$. Therefore overall execution time becomes $O(n^3 + m^3)$ time.

Proposition 6.4 *Graph-Alignment-Modified algorithm executes in $O(m_2.n^3)$ time.*

The *for* loop in line 2 of Algorithm 15 executes m_2 times, where $|E_2| = |m_2|$. The geometric-transform operation in line 3 takes $O(n_2^2)$ time; once a transformation matrix is constructed, where $|V_2| = n_2$. The EDM operation in line 4 is performed in $O(n^3)$ time, where $n = max(n_1, n_2)$. Hence total time taken by this algorithm is $O(m_2.n^3)$.

Algorithm 15 :Graph-Alignment-Modified(G_1, G_2)

Require: Two undirected geometric graphs G_1, G_2, where $G_1 = (V_1, E_1, c_1), |V_1| = n_1, |E_1| = m_1, G_2 = (V_2, E_2, c_2), |V_2| = n_2, |E_2| = |m_2|$

Ensure: Geometric transform of G_2 having maximum edges and vertices aligned to G_1

```
Select an edge e ∈ E_1 as a reference axis
for (each edge f ∈ E_2) do
    G'_2 ← geometric-transform(G_2)
    d ← EDM(G_1, G'_2)
    if (d < d_min) then
        d_min ← d
        G ← G'_2
    end if
end for
return (G)
```

6.5 Experimental Evaluation

6.5.1 Exact Geometric Graph Matching

To assess the effectiveness Geometric-Graph-Isomorphism algorithm, we apply it for exact graph matching. We use letter dataset of IAM graph database repository for the demonstration of exact graph matching [148]. Each node in letter graphs is labeled with an (x, y) coordinate to represent its location in the two-dimensional plane.

Depending on three *if* loop conditions in the Geometric-Graph-Isomorphism algorithm, there are three cases to be considered. In the first case, we take several geometric graphs, apply composite geometric transformation using the translation, rotation, scaling and their combination to these sample graphs and verify whether the concerned pair is geometrically isomorphic using Geometric-Graph-Isomorphism algorithm. In the second case, we apply random distortion within a specified range to the vertices of the sample graphs and check whether they are t-tolerant isomorphic with respect to each other. Finally, in the third case, we apply the proposed algorithm to low, medium, and high distortions of letter dataset to compute the graph matching, and compare their relative graph distances.

In the first set of experiments, we take letter graphs from each of low, medium and high class and apply an arbitrary transformation to the coordinate of vertices. We then apply the Geometric-Graph-Isomorphism algorithm to each pair of the original and transformed graph to verify the isomorphism. Since the Graph-Alignment algorithm makes the reference coordinate of two

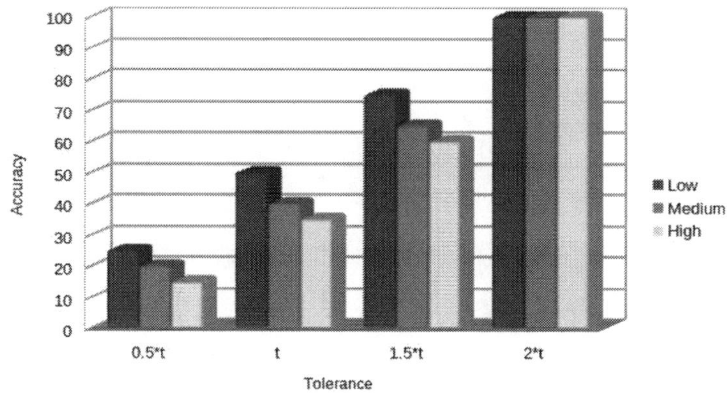

FIGURE 6.5: t-tolerant geometric graph isomorphism

graphs identical, it undoes the transformations applied on the processed letter graph to output the two graphs to be isomorphic in each test.

In the second set of experiments, we select letter graphs from each of low, medium, and high class, and apply random distortions from 0 to t distance on each coordinate of vertices of the graph. We check t-tolerant isomorphism for each pair of graphs for four different values of t which are $t/2, t, 3t/2, 2t$. Comparison of accuracy of t-tolerant isomorphism for different values of t, when Geometric-Graph-Isomorphism algorithm is applied to letter graphs of low, medium, and high distortions classes is shown in Figure 6.5. Here we observe that for the value of $2t$ every pair of graphs is correctly identified to be isomorphic.

In the third set of experiments, we apply Geometric-Graph-Isomorphism algorithm on each letter dataset to perform exact graph matching. For this experiment, we select graphs of equal vertices and edges. Based on the GD value returned by the algorithm, we compute the classification accuracy of letter dataset using the nearest neighborhood classifier. Classification accuracy of each of the fifteen letters of letter graphs of low distortion class is shown in Figure 6.6.

Similarly, classification accuracy of each letters for medium and high-distortion class of letter dataset is shown in Figure 6.7 and Figure 6.8, respectively.

The comparison of classification accuracy of each of the low, medium, and high class of letter dataset is shown in Figure 6.9. Here we observe that the accuracy of letter dataset of the low class is higher than that of medium class, which in turn is higher than that of letter dataset of high distortion.

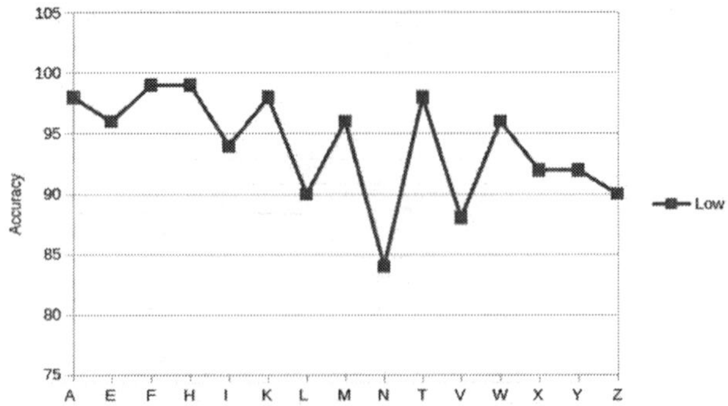

FIGURE 6.6: Accuracy low

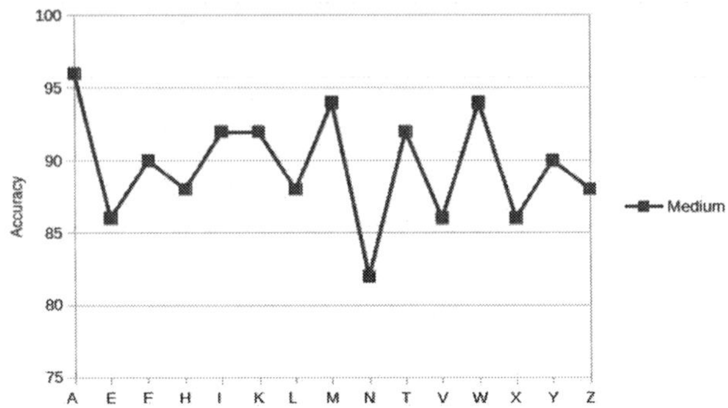

FIGURE 6.7: Accuracy Medium

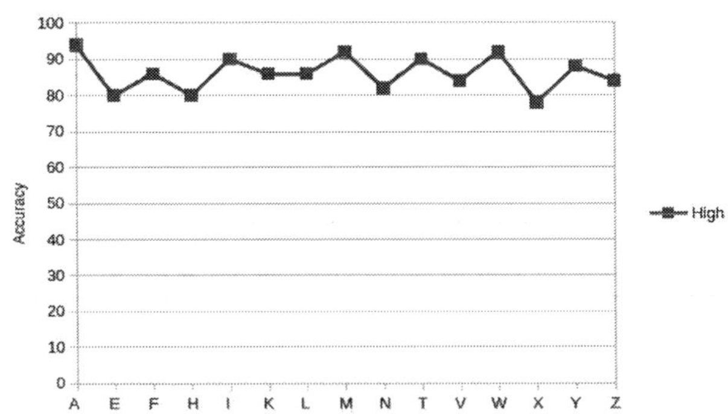

FIGURE 6.8: Accuracy High

6.5.2 Error-Tolerant Geometric Graph Matching

To evaluate the usefulness of the geometric graph similarity framework, we apply it for inexact graph matching task. In this section, we use

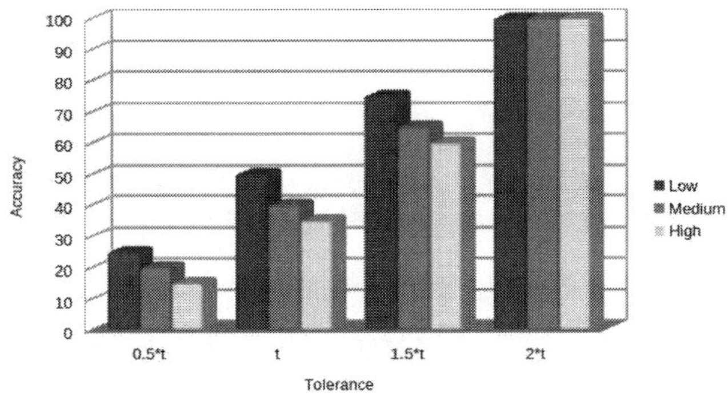

FIGURE 6.9: Accuracy

TABLE 6.1: Comparison of average computation time in ms

Algorithm	T_{Letter}	Algorithm	T_{AIDS}
Geometric	5.7	Geometric	1.39
Beam	4.5	Beam	6.48
GED	8.1	Bipartite	12.11

Geometric-Graph-Distance algorithm for geometric graph matching. We use the letter dataset and AIDS dataset of IAM graph database repository [148] for comparison of execution time and accuracy of geometric graph matching with other important graph matching algorithms. Each node in letter graphs is labeled with an (x, y) coordinate to represent its location in the two-dimensional plane. Every node of AIDS graphs also has its associated coordinate position.

6.5.2.1 Execution Time Comparison

Results in this section are computed using a system having 9.8 GB of memory and running the processor at 3.40 GHz. For comparison purpose, we use A^* based graph edit distance, beamsearch [139] with beam width $s = 10$, and bipartite graph matching [149]. Comparison of the average running time of graph matching methods in milliseconds for geometric graph matching, beamsearch having beam width $w = 10$, and GED is shown in Table 6.1.

We may not use A^* based graph edit distance on AIDS dataset, as it usually becomes infeasible beyond 10 to 20 nodes. Comparison of the average execution time of graph matching in milliseconds for geometric graph matching, beamsearch and bipartite graph matching is shown in Table 6.1. On an average geometric graph matching takes less execution time than either beamsearch optimization or bipartite graph matching heuristic.

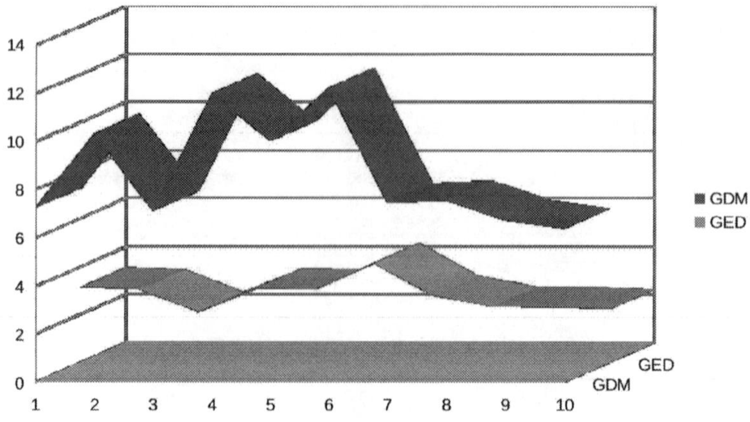

FIGURE 6.10: GDM vs GED for high-distortion letter graphs

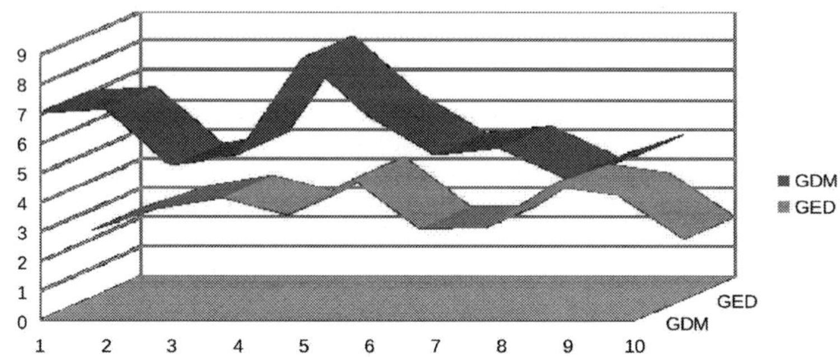

FIGURE 6.11: GDM vs GED for medium-distortion letter graphs

6.5.2.2 Accuracy Comparison

To compare the *GDM* computed by Algorithm 14 to the GED for the same pair of graphs, we compute these values for the first few letter graphs from each of high, medium, and low distortion letter graphs as shown in Figure 6.10, Figure 6.11, and Figure 6.12, respectively. We note that *GDM* and GED values are more similar in low distortion letter graphs differing only by a shift as compared to high-distortion letter graphs. It implies that in lower distortion, the two distance measures seems to correlate more than that with higher distortion levels. Also, the accuracy for low distortion letter dataset is higher than the accuracy for high-distortion dataset. It shows that the proposed model is more accurate for unmodified graphs.

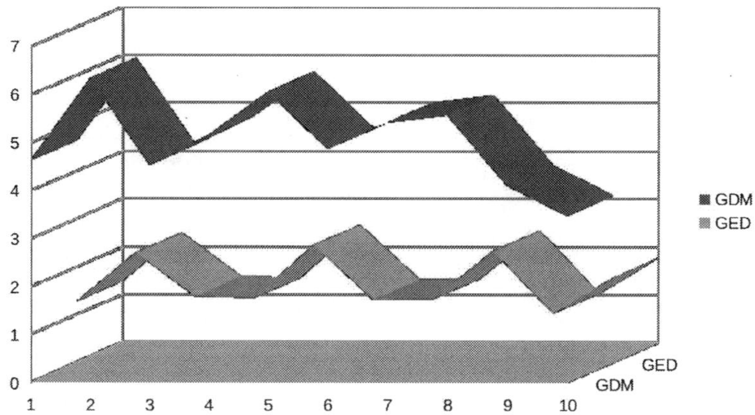

FIGURE 6.12: GDM vs GED for low-distortion letter graphs

For accuracy evaluation of geometric graph matching, we consider the problem of classification of letter graphs and AIDS graphs using nearest neighbor classifier. For accuracy evaluation of letter dataset, we use high-distortion letter graphs consisting of 50 training, validation and test graphs for each of the 15 classes of letters. We have trained the proposed model using the training set, optimized the weight parameters using the validation set and computed the classification accuracy using the test set. We use steepest-ascent search in the parameter space to find the weight parameters that are as well performing on the validation set as possible. The best weight parameter obtained are $w_1 = 0.35$, $w_2 = 0.23$, $w_3 = 0.11$, and $w_4 = 0.31$.

Table 6.2 shows the accuracy ratio for each of the 15 letters of high-distortion letter graphs using Algorithm 14. For comparison we have shown accuracy using VD, ED and EDM separately along with GD and GDM computed using Algorithm 14. We observe that accuracy using GDM is always higher than that of VD and EDM individually; similarly, the accuracy using GD is more than that of VD and ED separately. We also note that the average accuracy of GDM is higher than that of GD for the computation of graph distance in Algorithm 14.

To compare the accuracy of geometric graph matching with other important graph matching techniques, we apply these algorithms to letter dataset with high distortion. The average classification accuracy achieved using geometric GDM, beamsearch and GED are 89.06, 84.40, and 90, respectively. Comparison of the accuracy of letter dataset using geometric GDM, beamsearch and, GED for every individual letter is shown in Figure 6.13.

AIDS training dataset consists of 250 graphs, out of which 50 graphs are from active class and the remaining 200 from inactive class. Test dataset includes 1500 graphs, having 300 graphs from active and rest 1200 graph are from inactive class. The classification accuracy of AIDS dataset using

TABLE 6.2: Accuracy on letter dataset using VD, ED, EDM, GD, and GDM

Class	VD	ED	EDM	GD	GDM
A	84	80	84	98	98
E	36	78	80	80	82
F	72	64	70	86	88
H	38	72	74	82	84
I	96	76	78	96	94
K	42	74	82	92	94
L	84	52	60	88	90
M	38	88	88	96	96
N	34	64	68	84	84
T	84	62	74	90	92
V	74	50	58	88	88
W	38	82	82	94	96
X	30	52	66	70	72
Y	46	56	64	84	88
Z	40	68	72	88	90

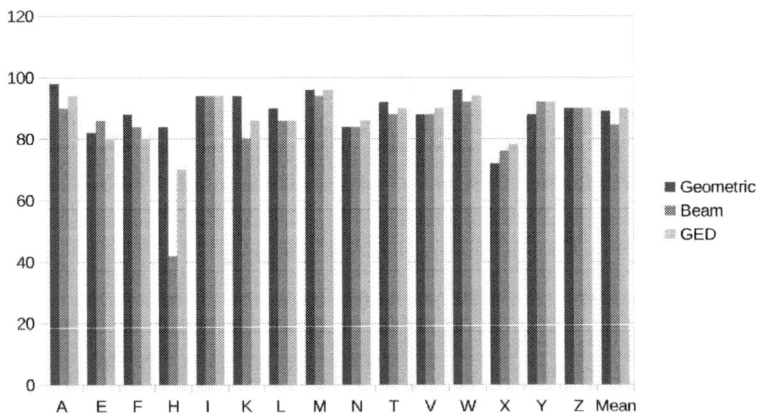

FIGURE 6.13: Comparison of accuracy on letter dataset

geometric GM, beamsearch and bipartite GM is shown in Table 6.3. We can observe that recognition rate of confirmed active molecules using geometric GM is more than both of beamsearch and bipartite GM, whereas classification rate of inactive molecules is less than that of beamsearch and bipartite GM.

TABLE 6.3: Accuracy on AIDS dataset

Algorithm used	Active	Inactive
Geometric	99.33	98.50
Beam	98.33	99.91
Bipartite	98.66	99.91

The proposed graph distance measure can be used to compare the structural similarity between different graphs. It can be particularly useful in real-world applications, where the graph data is large and can be modified by noise. Depending on application requirement, we can select weighting factors such that $\sum_{i=1}^{n} w_i = 1$. When the position of vertices is more dominant, we can select w_1 to be higher, if angular structures are more important then w_2 can be prominent. Otherwise, if edge differences are more essential, we can select w_3 to be higher. We can either use equal weight factors or train them to optimize the classification ratio. In this section, we have trained the weight parameters for the computation of GD and GDM in the Geometric-Graph-Distance algorithm.

6.6 Summary

In this chapter, an intuitive approach is described to compute similarity between two geometric graphs and used it to perform exact and error-tolerant graph matching. In a geometric graph, every vertex has an associated coordinate, which specify its distinct position in the plane. We use this fact to define the *VD* between the two graphs. We defined *EDM* between the two graphs using the angular, edge length, and position components between corresponding edges of two graphs. Finally, *VD* and *EDM* are combined to form *GDM* and used it to compute error-tolerant graph matching between two graphs. The experimental results show that this graph matching approach can be promising to graph dataset in which every node has a coordinate position in a two-dimensional plane.

Chapter 7

Graph Kernels and Embedding

Graph structures are pervasive in many domains, including biology, chemistry, social networks, and computer science. Analyzing these graphs requires powerful tools to understand their structure, similarity, and properties. Graph kernels and embeddings are two such tools that have become essential in the field of graph-based machine learning. This chapter explores these concepts, their mathematical underpinnings, and their applications.

A kernel is a function that computes the dot product in some feature space, allowing the comparison of two objects, such as graphs, without explicitly mapping them to that space. For graphs, a kernel function $K(G_1, G_2)$ computes the similarity between two graphs, G_1 and G_2. Using kernel functions, basic linear algorithms can be seamlessly extended to address complex non-linear problems in a unified framework. Additionally, kernel functions enable the application of standard algorithms to complex data structures that lack a conventional vector space representation.

Kernel machines are a class of algorithms in machine learning that rely on the concept of a kernel function to implicitly map input data into a high-dimensional feature space, where linear methods can be applied to solve complex problems. Theoretical analysis suggest that kernel machines, under certain conditions, are better suited for challenging pattern recognition tasks compared to traditional methods like nearest-neighbor classifiers or artificial neural networks. The key concept of kernel machines is kernel trick. The kernel trick allows these algorithms to operate in the original input space without explicitly computing the coordinates of data in the high-dimensional feature space. Instead, it computes the inner products of data points in the feature space directly using a kernel function.

Kernel machines have garnered significant attention in the fields of pattern recognition and machine learning, primarily due to the elegance of the kernel approach, its broad applicability to both standard and novel algorithms, and the robust theoretical foundation provided by statistical learning theory. This theory frames the task of learning a classifier from a training set in precise statistical terms, enabling the derivation of classifiers that are expected to perform effectively on unseen data.

Graph kernels are a class of functions designed to measure the similarity between graphs. They provide a way to map graphs into a feature space where traditional machine learning algorithms, such as Support Vector Machines (SVMs), can be applied.

DOI: 10.1201/9781003532248-7

This chpater is organized as follows. Section 7.1 explains the basic concepts and foundation of graph kernels. It describes various kernel functions. Section 7.2 presents important kernel machines including support vector machine, kernel principal component analysis, and kernel Fischer discriminant analysis. Section 7.3 describes various graph kernels. Section 7.4 describes Graph embedding.

7.1 Basic Concepts and Foundation

The prime concern of supervised learning is to derive from a finite training set of patterns and their labels $\{(x_i, y_i)\}_{i=1...m} \subseteq \mathcal{X} \times \mathcal{Y}$ a prediction function $f : \mathcal{X} \to \mathcal{Y}$ which assigns patterns $x \in \mathcal{X}$ to classes $y \in \mathcal{Y}$, that is, a function classifying patterns. When such a prediction function, or classifier, is provided, unknown patterns $x' \in \mathcal{X}$ can then be assigned to the class $f(x') \in \mathcal{Y}$. Here, the overall objective is to derive classifiers that are able to correctly classify not only the patterns from the training set $\{(x_i)\}_{i=1...m}$, but also other unseen patterns from the underlying population $\mathcal{X}$. This property is usually referred to as generalization ability of classifiers.

The supervised learning problem is usually reformulated in the form of the underlying distributions of patterns and labels for quantification of the generalization ability of various classifiers [166]. Under such conditions joint probability distribution measure $P(x, y)$ on $\mathcal{X} \times \mathcal{Y}$ is used to relate pattern with its classes. The probability distribution is usually not known and its remains unchanged. This model only assumes that the patterns and their class labels are selected randomly from a probability distribution. The classifier learning task is to derive a prediction function randomly selected from training patterns which can perform very well for the whole set of patterns.

The primary goal of supervised learning is not merely to encapsulate or memorize the training data with high accuracy, but rather to discern and model the fundamental characteristics of the underlying classes. This distinction is crucial because achieving a perfect fit to the training data does not necessarily imply good performance on unseen data. The ability of a model to generalize effectively to new data points drawn from the same distribution as the training data is the hallmark of successful supervised learning.

One of the major challenges in supervised learning is the phenomenon known as overfitting. Overfitting occurs when a classifier becomes excessively adapted to the peculiarity of the training set, capturing noise or spurious patterns instead of generalizing the characteristics of the underlying classes. Such a model may exhibit excellent performance on the training data but fails to generalize to new, unseen samples.

A significant focus in the design and development of machine learning algorithms has been on mitigating overfitting. Numerous strategies have been developed to address this issue. Architectural Simplification is used to reduce the number of neurons in hidden layers to decrease the model's capacity and prevent it from overfitting to small variations in the training set. Early Stopping monitors the model performance on a validation set during training and halting the training process before full convergence to avoid overadaptation to the training data. Regularization is another technique which introduces penalties in the loss function, such as L_1 or L_2 regularization, to constrain the model's parameter magnitudes. Dropout method is used to temporarily remove a random subset of neurons during training to prevent co-adaptation and encourage robustness.

In contrast, underfitting arises when a model fails to capture the complexity of the class boundaries and relationships within the data. This typically happens when the model's hypothesis class or capacity is insufficient to represent the underlying structure of the data. For instance, if a linear classifier is used on a dataset with nonlinear decision boundaries, the model will likely struggle to make accurate predictions.

Underfitting is characterized by both poor performance on the training set and an inability to generalize to new data. Common causes of underfitting include: overly simple models, insufficient training and indadequate feature representation. Using models with insufficient flexibility or too few parameters leads to overly simple models. Insufficient training does not allow the model to learn for a sufficient number of iterations, resulting in suboptimal parameter estimates. Failing to provide the model with informative features that capture the complexities of the problem domain results in indadequate feature representation.

Achieving a balance between overfitting and underfitting is a fundamental challenge in supervised learning. Models must be sufficiently complex to capture the underlying patterns in the data but not so complex that they adapt to noise or non-generalizable details in the training set. This balance can be achieved through careful selection of model architecture, regularization techniques, and hyperparameter tuning often guided by performance metrics on validation sets. While the overall goal of supervised learning is to capture the essence of the underlying classes, ensuring that a model generalizes well to unseen data requires addressing the twin challenges of overfitting and underfitting. By understanding these issues and employing appropriate strategies, machine learning practitioners can develop robust models capable of delivering accurate and reliable predictions in real-world scenarios.

7.1.1 Structural Risk Minimization

A key challenge in machine learning is finding a prediction function that strikes an optimal balance between underfitting and overfitting. This requires designing a classifier that is neither too simple nor overly complex, but instead as complex as necessary to accurately model the underlying class boundaries in the data. The principle of structural risk minimization (SRM) provides a formal framework for achieving this balance. The central idea of SRM is to quantitatively assess the degree of underfitting and overfitting for a given classifier and identify the model that is most likely to generalize effectively to unseen data. This involves selecting a model based on its complexity while ensuring it captures the true patterns in the data without being excessively flexible.

The complexity of a classifier is formally defined by its **capacity**, a measure of its ability to fit any training set without error. Capacity serves as an indicator of a model's flexibility or complexity:

- **Low-capacity classifiers** are simple models that may struggle to fit complex datasets, leading to underfitting.
- **High-capacity classifiers**, on the other hand, are capable of fitting even the most intricate training data, potentially memorizing noise and outliers, which can result in overfitting.

To illustrate, consider a dataset with three data points. There are $2^3 = 8$ possible ways to assign these points to two classes. A classifier with high capacity is one that can perfectly learn and classify all these eight labeling configurations. However, while such a classifier may achieve flawless performance on the training set, it might fail to generalize to new data points, as it lacks the ability to distinguish meaningful patterns from noise.

One of the most widely used measures of a classifier's capacity is the **Vapnik-Chervonenkis (VC) dimension**, introduced by Vapnik [190]. The VC dimension provides a theoretical foundation for evaluating a classifier's ability to generalize. Suppose a classifier is given m data points, each of which can be labeled with one of two class labels (e.g., -1 or $+1$). If the classifier can correctly learn all $2m$ possible label assignments for these m points, it is said to shatter the m points. The **VC dimension** of the classifier is defined as the largest m for which there exists a set of m points that the classifier can shatter. A *high VC dimension* indicates a model with high capacity, capable of fitting complex data patterns. However, such models are prone to overfitting. A *low VC dimension* corresponds to a simpler model with limited flexibility, which may underfit the data. The ideal classifier has a VC dimension that balances the ability to fit the training data while maintaining good generalization to unseen data.

The structural risk minimization principle leverages the concept of capacity and VC dimension to guide model selection. It aims to minimize the structural risk, which is a trade-off between two components: empirical risk and capacity complexity. While *empirical risk* is the error observed on the training data, *capacity complexity* is a penalty for excessive model complexity, quantified by measures such as the VC dimension. By optimizing this trade-off, SRM identifies a classifier with sufficient capacity to capture the essential characteristics of the data without overfitting. The interplay between underfitting and overfitting lies at the heart of supervised learning. The principle of structural risk minimization, supported by capacity measures such as the VC dimension, provides a systematic approach to selecting models that generalize well. This theoretical foundation underpins the design of robust learning algorithms, enabling practitioners to build classifiers that perform reliably on both training and unseen data.

It is important to recognize that numerous formal concepts exist for measuring the capacity of classifiers, as discussed in works such as [166] and [173]. Statistical learning theory emphasizes the necessity of restricting the set of functions from which a classifier is chosen. Specifically, the theory recommends selecting classifiers from function classes with capacities that are appropriate for the size of the available training dataset. While ensuring that a classifier correctly labels most training examples remains significant, it is neither necessary nor desirable for the classifier to achieve perfect accuracy on the training data. Overemphasis on fitting every training point, including outliers, can compromise the model's ability to generalize to new, unseen data. By limiting the capacity of a classifier, the influence of outliers in the training data is diminished. This adjustment often leads to better classification performance on data sampled from the same underlying distribution. Balancing classifier capacity with the size and complexity of the training data is a critical aspect of designing models that generalize effectively, as it mitigates overfitting while maintaining sufficient flexibility to capture meaningful patterns.

7.1.2 Kernel Function

Kernel functions are fundamental tools in pattern recognition and machine learning. They enable the construction of non-linear extensions to linear algorithms by embedding patterns into a high-dimensional feature space, where linear relationships can effectively capture non-linear structures in the original input space. This section introduces the concept of kernel functions, their mathematical properties, and their role in pattern recognition tasks.

Kernel functions serve as similarity measures between pairs of patterns. Unlike conventional measures, kernels implicitly map patterns into a high-dimensional space, known as the **kernel feature space**, without requiring explicit computation of the mapping. This feature enables kernel-based algorithms to handle complex, non-linear relationships efficiently.

A kernel function $k(x_i, x_j)$ computes the inner product between two patterns x_i and x_j in the feature space, providing a geometrical interpretation that is central to many algorithms, including support vector machines (SVMs). The choice of kernel determines the nature of the feature space and the patterns' representation, making it a critical factor in achieving effective classification and regression.

For a kernel function to be valid, it must satisfy certain mathematical properties, including symmetry and positive definiteness. These properties ensure that the kernel function is both mathematically sound and practically useful for constructing kernel matrices, which are essential components of kernel-based algorithms.

Definition 7.1 *A kernel function $k : \mathcal{X} \times \mathcal{X} \to \mathbb{R}$ is positive definite if it satisfies the following conditions:*

1. *Symmetry: $k(x_i, x_j) = k(x_j, x_i)$ for all $x_i, x_j \in \mathcal{X}$.*
2. *Positive Definiteness: For any $m \in \mathbb{N}$ and any real coefficient $\{c_1, ..., c_m \subseteq \mathbb{R}\}$, the following inequality holds: $\sum_{i,j=1}^{m} c_i c_j k(x_i, x_j) \geq 0$, where $\{x_1, ..., x_m\} \subseteq \mathcal{X}$.*

The **kernel matrix** or gram matrix is constructed using a positive definite kernel function and patterns $\{x_1, ..., x_m\}$. Its entries are defined as $K_{ij} = k(x_i, x_j)$. A kernel matrix is positive definite if all its eigenvalues are non-negative. Alternatively, it is positive definite if there exists a matrix B such that $K = B^T B$, or if all its principal minors (determinants of square submatrices) are non-negative.

The property of positive definiteness is crucial for kernel methods because it ensures the kernel matrix satisfies the mathematical requirements of optimization algorithms, particularly those used in support vector machines and Gaussian processes. These properties provide a foundation for the theoretical guarantees and stability of kernel-based models.

In addition to positive definite kernels, a broader class of functions called conditionally positive definite kernels is often used. These kernels are useful when positive definite kernels are unavailable or when additional flexibility is needed.

Definition 7.2 *A kernel $k : \mathcal{X} \times \mathcal{X} \to \mathbb{R}$ is conditionally positive definite if, for any positive integer $m \geq 2$, the following inequality holds:*

$$\sum_{i,j=1}^{m} c_i c_j k(x_i, x_j) \geq 0$$

where $\{c_1, ..., c_m \subseteq \mathbb{R}\}$ satisfy $\Sigma_{i=1}^{m} c_i = 0$ and $\{x_1, ..., x_m\} \subseteq \mathcal{X}$.

Every positive definite kernel is conditionally positive definite, but not vice versa. Conditionally positive definite kernels can be applied in certain algorithms as substitutes for positive definite kernels, expanding the range of applicable problems.

Kernel functions can be composed or modified to create new valid kernels. For example, pointwise addition and pointwise multiplication operations preserve positive definiteness. These operations enable the design of complex kernels tailored to specific problem domains, combining the strengths of multiple kernel functions.

Kernel functions provide a powerful framework for extending linear algorithms to non-linear settings, enabling effective pattern recognition and machine learning in high-dimensional spaces. The requirements of symmetry and positive definiteness ensure their theoretical soundness, while the flexibility to construct derived kernels allows customization for diverse applications. By leveraging kernel functions, algorithms can achieve remarkable performance on complex datasets, making them an indispensable tool in modern machine learning.

Proposition 7.1 ***(Closure Properties)****. Let k_1 and k_2 be two positive definite kernels, and let $c_1, c_2 \geq 0$ and $b \in \mathbb{R}$. The following kernel functions exhibit specific closure properties:*

- $k_a(x, x') = c_1 k_1(x, x') + c_2 k_2(x, x')$ *(Sum of Kernels)*
- $k_b(x, x') = k_1(x, x').k_2(x, x')$ *(Product of Kernels)*
- $k_c(x, x') = k_1(x, x') + b$ *(Affine Transformation of a Kernel)*

Where kernels k_a and k_b are positive definite, whereas k_c is conditionally positive definite.

For a pattern recognition problem to be solved using kernel machines, one of the most challenging aspects is defining a kernel function that effectively captures the relevant information from the underlying patterns. When patterns are commonly represented as feature vectors, several standard valid kernels are widely employed. [166]

A vast array of linear methods for pattern analysis and classification has been developed to address fundamental tasks such as achieving linear separation of data or identifying dominant linear relationships within datasets. These methods are invaluable for gaining insights into and processing data. However, their utility is inherently constrained by their linear nature. When non-linear regularities arise, linear methods are inadequate and fail to capture the complexities of the data.

A universal approach to overcoming this limitation involves transforming the problem domain so that non-linear regularities are converted into linear ones. Mapping data into a higher-dimensional space that is better suited to the recognition task at hand is a particularly effective strategy. The core challenge in this context lies in defining the high-dimensional space into which patterns are mapped and determining the mapping function itself. Cover's theorem [47] provides a theoretical foundation for this approach. The theorem states that a set of patterns associated with a complex recognition problem, when cast non-linearly into a sufficiently high-dimensional space, is more likely to become linearly separable in that space compared to the original low-dimensional space. This implies that even when the precise properties of the high-dimensional feature space required to simplify the recognition problem are unknown, constructing such a space alongside a suitable non-linear mapping is statistically advantageous.

Many pattern analysis and classification algorithms are inherently defined using geometrical constructs. These include distances to prototype patterns, projections into lower-dimensional subspaces, linear combinations of patterns, and the convex hulls of pattern sets, among others. The ability to compute geometrical relationships in the feature space where the recognition problem is addressed is thus critical. Notably, certain geometrical relationships, such as how patterns are interconnected, can often be expressed through inner products. This is particularly true when the feature space itself is a vector space, underscoring the importance of inner-product-based computations for effective pattern recognition in higher-dimensional spaces.

Inner products can be interpreted as measures of pattern similarity. For example, the standard inner product between two vectors is maximized when the vectors are aligned in the same direction and minimized when they are oriented in opposite directions. Interestingly, the inner product in a higher-dimensional feature space H can be inferred from the inner product in the original space X, without the need to explicitly map the patterns into H. While deriving the inner product of a three-dimensional space from a two-dimensional space might seem trivial, this concept extends to a much broader class of functions and applies to high-dimensional or even infinite-dimensional vector spaces.

Theorem 3 ***(Kernel Trick)*** *Consider a valid kernel function $k : \mathcal{X} \times \mathcal{X} \rightarrow \mathbb{R}$. There exists a vector space $\mathcal{H}$, equipped with an inner product $\langle .,. \rangle : \mathcal{H} \times \mathcal{H} \rightarrow \mathbb{R}$, and a mapping $\phi : \mathcal{X} \rightarrow \mathcal{H}$ such that:*

$$k(x, x') = \langle \phi(x), \phi(x') \rangle$$

for all $x, x' \in \mathcal{X}$.

Even though this theorem may appear straightforward, its implications for pattern recognition methods are profound. Instead of explicitly mapping patterns from $\mathcal{X}$ to $\mathcal{H}$ and computing their inner product in $\mathcal{H}$, one can directly compute the kernel function $k(x, x')$ in $\mathcal{X}$. Conversely, every kernel function can be interpreted as an inner product in an explicitly existing feature space.

This condition of positive definiteness (and symmetry) guarantees that any valid kernel function represents a rich structure of vectors and geometric entities, even when the original pattern space $\mathcal{X}$ lacks intrinsic mathematical structure. This property enables the application of well-established algorithms designed for vector spaces to non-vectorial data in a unified manner. For instance, while it is impossible to compute the average of a set of graphs in the graph space, it becomes straightforward to compute the average of vectors representing those graphs in the feature space. The structure of this feature space $\mathcal{H}$ is entirely determined by the kernel function, which serves as a similarity measure in $\mathcal{X}$. To embed $\mathcal{X}$ into $\mathcal{H}$, one only needs to define a symmetric and positive definite similarity function on $\mathcal{X}$, which subsequently dictates the geometry of $\mathcal{H}$.

This kernel-based methodology provides a powerful framework for transforming algorithms. Any algorithm formulated solely in terms of an inner product—or more generally, a symmetric positive definite function—can be adapted by replacing the inner product with a valid kernel function. This procedure, known as the **kernel trick** [166],[173], allows algorithms to operate in an implicitly defined feature space without explicitly constructing the vectors or calculating inner products.

In practical pattern recognition, the kernel trick plays a crucial role, as it facilitates the extension of linear algorithms to handle non-linear problems with minimal modification. Algorithms inherently reliant on inner products, such as those for linear operations, can be seamlessly adapted to exploit the kernel framework. The kernel trick is applicable to a wide range of important algorithms, including support vector machines (SVMs), principal component analysis (PCA), Fisher discriminant analysis (FDA), nearest-neighbor classifiers, k-means clustering, self-organizing maps, and many others. By leveraging this approach, one can address complex, non-linear pattern recognition tasks with remarkable efficiency and flexibility.

7.2 Kernel Machines

This section introduces the widely used kernel machines such as support vector machines, kernel principal component analysis, and kernel Fisher discriminant analysis. Among these, the support vector machine is particularly notable for its popularity in pattern classification tasks. Its appeal lies in a training process grounded in statistical learning theory and its consistently impressive performance compared to traditional classifiers across diverse datasets. In contrast, kernel principal component analysis and kernel Fisher discriminant analysis serve as non-linear extensions of classical feature extraction techniques, offering enhanced capabilities for handling complex data structures.

7.2.1 Support Vector Machine

Support Vector Machines (SVMs) are supervised learning algorithms used for classification and regression tasks. The core idea of SVMs is to find the optimal hyperplane that separates data points belonging to different classes with the largest possible margin. In essence, SVMs aim to maximize the margin, which is the distance between the hyperplane and the nearest data points from each class, known as support vectors.

In cases where data is not linearly separable, SVMs utilize kernel functions to map the input data into a higher-dimensional space where linear separation becomes feasible. This mapping allows SVMs to handle both linear and non-linear classification problems effectively.

The basic idea of SVM is based on hyperplane, margin and support vectors. *Hyperplane* is a decision boundary the separates data points of different classes. *Margin* is the distance between the hyperplane and the closest data points from each class. SVM maximizes this margin to improve generalization. *Support vectors* are the data points closest to the hyperplane, which determine its position and orientation.

Let's consider a simple example where we want to classify two types of data points: white circles and black squares.

- Dataset: Assume we have a dataset of points in a two-dimensional space. White circles are located near (1,2), (2,3), and (3,1). Black squares are located near (6,5), (7,6), and (5,7).
- Linear Separability: In this case, the data points can be separated by a straight line (hyperplane). A potential hyperplane could be: $y = (1/2)x + 2$. However, SVM does not just find any separating line; it identifies the optimal hyperplane that maximizes the margin.
- Support Vectors: The points closest to the hyperplane on both sides are the support vectors. These points are critical because they define the hyperplane.
- Maximizing the Margin: SVM calculates the hyperplane that ensures the maximum margin between the support vectors of the two classes. This results in better generalization to new, unseen data.

If the data cannot be separated by a straight line (e.g., white circles form a ring around black squares), SVM uses a kernel function to map the data into a higher-dimensional space where the separation is linear. For instance the Radial Basis Function (RBF) Kernel maps the data points into a higher-dimensional space and separates them using a hyperplane in that space.

The decision function of SVM is defined as:

$$f(x) = \sum_{i=1}^{n} \alpha_i y_i k(x_i, x) + b$$

Where, x_i is training data points, y_i represent labels, α_i denote Lagrange multipliers, $k(x_i, x)$ is kernel function and b is bias term.

As a practical example, imagine a spam email classification problem. Where input is the dataset of emails represented as feature vectors (e.g., word frequencies, presence of keywords). Output is class of spam and non-spam email. SVM identifies an optimal hyperplane in the feature space that separates spam emails from non-spam emails. If non-linear relationships exist (e.g., combinations of words indicating spam), a kernel function is used to capture these complexities. Once trained, the SVM can classify new emails as spam or not spam based on which side of the hyperplane they fall.

Primary advantage of SVM is accuracy; by maximizing the margin, SVM ensures robust classification. It is also flexible as kernel functions allow handling non-linear relationships in the data. Another advantage is generalization since the model performs well on unseen data by focusing on support vectors and ignoring outliers. SVM's ability to handle both simple and complex patterns makes it a highly effective tool for tasks like email filtering, image recognition, and bioinformatics.

7.2.2 Kernel Principal Component Analysis

Kernel Principal Component Analysis (KPCA) is a non-linear extension of the traditional Principal Component Analysis (PCA). It is used for dimensionality reduction and feature extraction in datasets where the relationships between features are non-linear. Unlike PCA, which identifies linear components in the data, KPCA leverages kernel functions to map the data into a higher-dimensional space, where linear relationships can be identified, effectively capturing non-linear patterns in the original data.

KPCA begins by mapping the data from its original space $\mathcal{X}$ into a higher-dimensional feature space $\mathcal{H}$ using a non-linear mapping function $\Phi : \mathcal{X} \rightarrow \mathcal{H}$. However Φ is not explicitly computed. Instead, kernel functions are used to compute dot products in $\mathcal{H}$, which are sufficient for the algorithm. In the feature space $\mathcal{H}$, the pairwise dot products of the mapped data points are represented using a kernel function . These products form the kernel matrix, K which ia central to KPCA. Since PCA requires the data to be centered (mean subtracted), KPCA adjusts the kernel matrix to ensure the mapped data in $\mathcal{H}$ is also centered. The centered kernel matrix is subjected to eigenvalue decomposition to obtain eigenvalues and eigenvectors. Here eigenvalues represent the variance explained by each principal component whereas eigenvectors correspond to the principal direction in the feature space $\mathcal{H}$. Finally, data is projected onto the top principal components (eigenvectors associated with the largest eigenvalues), which capture the most significant patterns in the data.

Suppose we have a dataset of points arranged in a circular pattern, which PCA fails to handle because the data's variance is non-linear. KPCA, using an RBF kernel, maps the data into a higher-dimensional space where the circular pattern is unfolded into a linear one. This enables the identification of principal components that capture the inherent structure of the data, facilitating effective dimensionality reduction or feature extraction.

Advantage of KPCA is that it allows to capture non-linear relationships that traditional PCA cannot handle. Also, by transforming the data into a feature space where non-linear patterns become more distinguishable, KPCA improves the effectiveness of downstream machine learning algorithms. The kernel trick enables KPCA to compute high-dimensional representations without explicitly performing expensive computations, reducing computational overhead. KPCA allows the use of different kernel functions making it adaptable to various types of data distributions. KPCA is particularly useful when the original feature space is complex or lacks meaningful structure, as it can reveal hidden patterns in high-dimensional spaces. Its flexibility, effectiveness in high-dimensional spaces, and ability to capture complex data structures make it a valuable tool in pattern recognition and machine learning applications.

7.2.3 Kernel Fisher Discriminant Analysis

Kernel Fisher Discriminant Analysis (KFDA), also known as Kernel Fisher Discriminants (KFD), is a non-linear extension of Linear Discriminant Analysis (LDA). KFDA combines the principles of discriminative analysis with kernel methods to handle non-linear separability in data. Like LDA, it aims to find a projection that maximizes the class separability, but it achieves this in a high-dimensional feature space induced by a kernel function.

One of the primary limitations of KPCA is that it is an unsupervised method, meaning it does not utilize class label information when performing dimensionality reduction. While KPCA effectively identifies directions of maximum variance in the data, it does not ensure that these directions are optimal for distinguishing between different classes. Consequently, KPCA may not always be suitable for classification tasks where class separability is crucial.

Kernel Fisher Discriminant Analysis offers a supervised alternative to KPCA and can be viewed as a method that incorporates class information during feature extraction. Unlike KPCA, which focuses solely on preserving variance, KFDA explicitly seeks to maximize inter-class separation while minimizing intra-class variability. This means that KFDA finds a transformation that ensures that different classes remain well-separated while data points belonging to the same class form compact clusters in the transformed space. [55]

To achieve this, KFDA projects the input data onto a lower-dimensional subspace in such a way that the distance between the means of different classes is maximized while the variance within each class is minimized. This optimization process is formulated as an eigenvalue problem, similar to KPCA. However, unlike KPCA, which does not consider class labels, KFDA takes advantage of supervised learning principles, making it more effective for classification tasks.

Since KFDA ensures better class separability, classification in the KFDA-transformed space is significantly easier compared to KPCA. In complex pattern recognition problems, ignoring class label information—as KPCA does—can lead to suboptimal results. As a result, KFDA is often preferred over KPCA for supervised learning tasks where discriminative feature extraction is essential.

7.3 Graph Kernel

Using kernel machines to address the graph matching problem introduces a transformative perspective compared to traditional approaches. The key advantage lies in the formulation based on kernel functions, which enable the embedding of graphs into a structured vector space through graph kernel similarity measures. This embedding facilitates a seamless representation of the space of graphs within an inner product space, allowing for advanced computational techniques to be applied.

Graph kernels play a central role in this process by defining a similarity measure between graphs that satisfies specific mathematical properties. By leveraging such a similarity measure, we can effectively map the space of graphs into a corresponding vector space. This mapping resolves a fundamental challenge in graph analysis: the inherent lack of mathematical structure in the original space of graphs. For instance, operations as basic as calculating the average of multiple graph patterns are not straightforward within the space of graphs. However, these operations become trivial in a vector space, where well-established mathematical tools can be utilized.

The benefits of this approach extend beyond simplification. Algorithms designed to operate in vector spaces—relying on concepts such as distances, lengths, angles, projections, and linear combinations—can be directly applied to graph matching tasks without the need for laborious adaptations. In contrast, performing such computations directly in the graph domain requires a comprehensive and often intricate redefinition of these quantities, making the process cumbersome and time-intensive.

By embedding graphs into a vector space using kernel functions, we circumvent these challenges entirely. The vector space provides a robust mathematical framework where basic operations and advanced manipulations are inherently supported. Consequently, the task of graph matching is transformed from a complex combinatorial problem into one that can leverage efficient, well-established techniques in vector spaces. This approach not only enhances computational efficiency but also broadens the scope of methods that can be applied to graph-related tasks, offering a powerful toolkit for addressing diverse problems in graph matching and analysis.

7.3.1 Model Driven Kernels

Model-driven kernels leverage domain-specific knowledge about the sample space, particularly the relationships and dependencies among the data points. This approach provides a structured framework for deriving meaningful features or transformations that capture the intrinsic properties of the data. Broadly, model-driven kernels can be categorized into two main types: generative models and transformative models. Each of these approaches offers a unique way of extracting insights from the data and constructing kernels tailored to specific problem domains.

Generative Models: In generative models, the parameters of the model are treated as features that represent the data. These parameters are derived from probabilistic or statistical models designed to describe how the data is generated. By focusing on the underlying process that gives rise to the observed data, generative models enable the creation of kernels that compare data points based on their likelihood or their fit to the model. For instance, in the context of graph kernels, the structural features of graphs, such as node distributions, edge weights, or clustering coefficients, can be incorporated into the generative framework, capturing both local and global properties of the graph. This approach is particularly effective in domains where the generative process is well understood and can be modeled accurately.

Transformative Models: Transformative models, on the other hand, focus on the ability of the data—such as graphs in the context of graph kernels—to undergo specific transformations that are relevant to the problem domain. These transformations alter the graph in a way that highlights its structural or relational properties. The transformed graphs are then treated as representations of the instance space, encapsulating the key characteristics of the data in a form suitable for kernel computation. A critical aspect of transformative models is their ability to capture both local and global information within the data. During the transformation process, individual edges primarily encode local information about neighboring vertices, reflecting the immediate relationships within the graph. However, the collection of all edges across the transformed graph provides a more holistic view, encapsulating the global structure and connectivity patterns of the sample space. This dual representation ensures that the kernel can effectively differentiate between graphs based on both granular and overarching structural features.

Unified Perspective: Model-driven kernels, whether generative or transformative, offer a powerful way to incorporate prior knowledge and structural insights into the kernel design process. By treating the sample space as a structured entity and explicitly modeling its properties, these kernels enable more accurate and meaningful comparisons between data points. This approach is especially valuable in applications such as bioinformatics, social network analysis, and pattern recognition, where the data often exhibits complex, interconnected structures that require sophisticated methods for analysis.

7.3.2 Syntax Driven Kernels

Syntax-based models prioritize the structural and semantic aspects of the data, offering a systematic way to analyze and compare complex patterns. These models represent one of the largest families of kernels, commonly referred to as convolutional kernels. Their foundation lies in the decomposition of structured data into smaller, syntactic elements, which are then used to define kernel functions that measure similarity or other relational properties.

Syntax-based models operate by breaking down structured objects, such as graphs or trees, into fundamental components. These components, often referred to as syntactic elements, include:

1. **Walks**: Sequences of consecutive vertices connected by edges within a graph.
2. **Paths**: Similar to walks but without repeated vertices, ensuring a unique traversal route.
3. **Cycles**: Closed walks where the starting and ending vertices are the same.
4. **Subgraphs**: Smaller graphs that are contained within a larger graph, preserving specific structural relationships.
5. **Subtrees**: Hierarchical structures that represent partial tree-like configurations within a dataset.

By focusing on these elements, convolutional kernels capture the intricate patterns that define the structure and relationships within the data.

One of the most compelling features of syntax-based models is their ability to preserve both local and global characteristics of the data. **Local Features** include syntactic elements such as walks or small subgraphs encapsulate fine-grained, localized information about the relationships between neighboring entities. For example, in a graph, local features might represent how individual nodes interact with their immediate surroundings. **Global Features** include elements such as longer paths, cycles, or larger subgraphs aggregate information from across the entire structure, offering insights into the overall topology and high-level organization of the data. This dual focus allows syntax-based kernels to provide a comprehensive representation of structured data, making them applicable to a wide range of tasks.

Due to their flexibility and robust feature preservation, syntax-based kernels find utility in numerous domains, including pattern recognition, social network analysis, natural language processing and bioinformatics. The ability of syntax-based models to handle diverse structural elements makes them a versatile and powerful tool. They are particularly effective in scenarios where the data exhibits rich interconnections and hierarchical relationships. Moreover, the convolutional nature of these kernels enables efficient computation, even when dealing with complex, high-dimensional data.

By focusing on the syntactic breakdown of structured objects and leveraging convolutional kernels, syntax-based models offer a framework that balances granularity with holistic analysis, making them indispensable in addressing a variety of real-world problems.

7.3.3 Convolution Kernel

A convolution kernel is a type of kernel function used to measure the similarity between structured data by decomposing it into smaller substructures, comparing these components, and aggregating the results. It is a powerful tool in machine learning, especially in domains where data is represented as graphs, trees, sequences, or other hierarchical structures. The term "convolution" here is inspired by the mathematical concept of convolution, referring to the process of combining information from smaller units to compute a result.

Convolution kernels work by breaking down complex objects (like graphs or strings) into smaller parts or substructures, such as nodes, edges, subtrees, walks, paths, or sequences. For example, a graph could be decomposed into its edges, subgraphs, or cycles. A string might be broken into substrings or n-grams. After decompostion of data the kernel function computes a similarity score between corresponding substructures from two data objects. This is often done using simpler kernel functions that operate on individual substructures. For instance, two edges in different graphs might be compared based on their weights, labels, or other properties. After comparing the substructures the similarities of the substructures are aggregated to compute an overall similarity score for the entire objects. This aggregation often involves summing or averaging the individual similarities. Convolution kernels can operate recursively, meaning they consider relationships between substructures at different levels. For instance, when comparing two graphs, the kernel might first compare their nodes, then edges, and finally larger subgraphs.

Definition 7.3 *Let x and x' represent two structured objects. A convolution kernel is defined as:*

$$k(x, x') = \sum_{(u,v) \in U(x) \times U(x')} k(u, v)$$

Where $U(x)$ and $U(x')$ are sets of substructures of x and x', respectively, $k(u, v)$ is a base kernel function that computes the similarity between substructures u and v and the summation aggregates the similarities across all pairs of substructures.

Convolution kernels capture both local features (through individual substructure comparisons) and global features (through aggregation over the entire structure). They can be applied to a wide variety of data types, including graphs, sequences, images, and text. Specific types of convolution kernels include graph kernels (e.g., random walk kernels, Weisfeiler-Lehman kernels) and string kernels. By focusing on substructures, convolution kernels are robust to noise or small changes in the data. For example, a small modification to a graph may not significantly alter the overall kernel value. Many convolution kernels are designed to compute similarity efficiently, even for large and complex data objects, often using recursive or dynamic programming techniques.

In a graph kernel, a convolution kernel might decompose graphs into subgraphs and compare their similarities:

- Decompose graph G_1 into subgraphs $\{S_1, S_2\}$.
- Decompose graph G_2 similarly into $\{T_1, T_2\}$.

- Compute pairwise similarities $k(S_i, T_j)$ between subgraphs from G_1 and G_2.
- Aggregate these similarities to get the overall kernel value $K(G_1, G_2)$

Convolution kernels thus provide a principled way to analyze and compare structured data efficiently while preserving meaningful relationships within the data.

7.3.4 Diffusion Kernel

The diffusion kernel is a kernel function used in pattern recognition and data analysis to measure the similarity between data points based on a diffusion process. It is particularly well-suited for data represented as graphs, where the relationships between nodes can be naturally modeled by the spread of information or influence over the graph structure. The diffusion kernel is grounded in principles from physics and probability, particularly the heat diffusion process.

Imagine placing a unit of heat (or a probability mass) at one point in a graph or data space. Over time, this heat diffuses through the structure, influenced by the connections between points. The diffusion kernel captures the similarity between points by considering how much heat or influence transfers between them during this process. Intuitively, points that are closely connected or highly similar will share more heat while points that are distant or weakly connected will share less heat.

The diffusion kernel is often derived from the Laplacian matrix of a graph, which encodes its structural properties. Let L denote the Laplacian matrix of the graph, and let t be a diffusion time parameter. The diffusion kernel is defined as:

$$K = e^{-tl}$$

Where, e^{-tl} is the matrix exponential of $-tl$, which can be computed as series expansion: $e^{-tL} = 1 - tL + (tL)^2/2! - (tL)^3/3! + (tL)^4/4! - ....$ t is a parameter that controls the extent of the diffusion. A smaller t considers local relationships, while a larger t incorporates global structure.

The diffusion kernel matrix K is symmetric and positive semi-definite, ensuring it can be used a valid kernel in machine learning algorithm. The diffusion process considers both local neighborhood structures and broader graph connectivity. The parameter t provides flexibility, allowing the kernel to focus on either short-range or long-range relationships. Advantage of diffusion kernel is its flexibility to adapt to various graph structures and scales. It exploits both local and global features of the graph. It also handles noise and irregularities in data due to smoothing effect of diffusion.

The diffusion kernel is widely used in domains that involve graph-based or relational data including graph matching, image processing and bioinformatics. One of the limitations of the diffusion kernel is its computational complexity as computing the matrix exponential e^{-tL} can be expensive for large graphs. Another limitation is its parameter sensitivity because choosing an appropriate value for t can significantly affect performance. The diffusion kernel remains a powerful and versatile tool for analyzing structured data, offering insights that leverage both the local and global properties of the underlying graph.

7.3.5 Random Walk Kernel

The Random Walk Kernel is a graph kernel used to measure the similarity between two graphs based on the random walks performed on them. It is a widely used method in pattern recognition and data mining tasks that involve structured data, such as graph classification, clustering, and regression. The kernel leverages the concept of random walks—sequences of steps through the nodes of a graph—to compare the structural and relational properties of graphs.

A random walk is a sequence of transitions between nodes in a graph, where each step is determined by the edges and their associated probabilities. The random walk kernel computes the similarity between two graphs by counting common random walks in both graphs. Intuitively, if two graphs have many similar random walks, they are likely to have similar structures also the length and nature of these walks encode both local (node-level) and global (graph-level) structural information.

Definition 7.4 ***(Product Graph)*** *Product graph of two graphs G_1 and G_2 is defined as $G_\times = V_\times, E_\times$, such that*
$V_\times(G_1 \times G_2) = \{(u_1, v_1) \in V_1 \times V_2 : label(u_1) = label(v_1)\}$,
$E_\times(G_1 \times G_2) = \{((u_1, v_1), (u_2, v_2)) \in V^2(G_1 \times G_2) : (u_1, u_2) \in E_1 \wedge (v_1, v_2) \in E_2 \wedge (label(u_1, u_2) = label(v_1, v_2))\}$

In other words product graphs of two graphs consists of pairs of identically labeled nodes and edges from G_1 and G_2.

Definition 7.5 ***(Random Walk Kernel)*** *Let G_1 and G_2 be two graphs. The random walk kernel of G_1 and G_2 is defined as:*

$$k_\times(G_1, G_2) = \sum_{i,j=1}^{|V_\times|} [\sum_{n=0}^{\infty} \lambda^n A_\times^n]_{ij}$$

Where $A_\times$ is the adjacency matrix of the product graph $G_1 \times G_2$, and λ is decaying factor.

The random walk kernel value reflects the weighted sum of the counts of matching random walks of various lengths between the two graphs. The random walk kernel is a powerful tool for graph similarity analysis, offering a balance between computational tractability and rich structural representation. By focusing on the shared sequences of transitions in graphs, it provides insights into their underlying topologies and relational patterns.

7.3.6 Similarity Kernel from Graph Edit Distance

The graph edit distance (GED) is a widely used and flexible measure of dissimilarity for graphs, defined as $d : G \times G \rightarrow R$. Given its effectiveness in quantifying how different two graphs are, it is natural to consider transforming this dissimilarity measure into a similarity measure. This approach offers the advantage of eliminating the need to define entirely new kernel functions specifically tailored for measuring graph similarity. Instead, the existing dissimilarity values are transformed into similarity values that can be directly utilized in kernel-based machine learning algorithms.

The transformation leverages a simple yet effective principle: low dissimilarity values correspond to high similarity values, and high dissimilarity values correspond to low similarity values. This is achieved using a monotonically decreasing function.

Definition 7.6 *For two graphs G_1 and G_2, the similarity kernel is defined as:*

$$k(G_1, G_2) = \exp^{(-\gamma d(G_1,G_2)^2)}$$

Where $\gamma > 0$ is a scaling parameter.

Above scaling parameter controls the sensitivity of the kernel to variations in the edit distance. The similarity kernel maps distances into similarity scores using an exponential decay function. This ensures that graphs with smaller edit distances are assigned higher similarity values. Since the kernel is derived directly from the edit distance, it can be applied to a wide range of problems without requiring domain-specific modifications. The parameter γ controls the sharpness of the similarity decay. Smaller values of γ result in a slower decay, making the kernel less sensitive to differences in the edit distance, while larger γ values emphasize even small differences.

This approach is particularly well-suited for distance-based data classification tasks. By transforming edit distances into similarity values, the resulting kernel can be directly fed into kernel-based algorithms such as SVMs. These algorithms rely on similarity measures to construct decision boundaries, making this transformation essential for leveraging the power of GED in such contexts.

One important limitation of the above similarity kernel is that it is not guaranteed to be positive definite. Positive definiteness is a desirable property for kernel functions, ensuring that the associated Gram matrix is valid for many kernel-based learning methods. However, theoretical research indicates that indefinite kernels can still be effective in practice under certain conditions.Empirical evidence also suggests that kernel machines can perform well even when the kernel is not strictly positive definite.

The transformation of the graph edit distance into a similarity kernel offers a practical and efficient way to leverage the dissimilarity measure for graph-based learning tasks. Despite the potential challenge of indefiniteness, this kernel provides a straightforward and flexible tool for comparing graphs in a variety of applications. By combining the expressiveness of the edit distance with the power of kernel-based methods, this approach bridges the gap between distance-based and similarity-based graph analysis.

7.4 Graph Embedding

Graph embedding techniques are methods that map graph-structured data into a lower-dimensional vector space while preserving the graph's structural and semantic properties. These embeddings transform nodes, edges, or entire subgraphs into fixed-length numerical vectors, enabling the application of machine learning and data analysis techniques that work with vector-based data.

Graphs are inherently complex, and operations like pattern matching, similarity measurement, or classification are computationally intensive in their raw form. By embedding graphs into a vector space allow the use of standard machine learning algorithms (e.g., clustering, classification, regression). Graph embeddings often reduce high-dimensional graph structures into more manageable representations also the well-designed embeddings retain the graph's key topological, structural, and relational properties.

The core idea behind the graph embedding approach is to represent a graph by its distances to a set of selected prototype graphs, creating a vectorial description of the input graph. Instead of working directly with the graph-based representation, this method employs the dissimilarity representation for pattern recognition, enabling the application of vector-based machine learning techniques [144].

Let us assume a sample set of graphs $\mathcal{S} = \{g_1, g_2, ..., g_N\}$ from a graph domain $\mathcal{G}$. Let $d : \mathcal{G} \times \mathcal{G} \to \mathbb{R}$ be a graph dissimilarity measure such as GED. The set $\mathcal{S}$ can consist of real-world graphs or even synthesized graphs for convenience. Typically, $\mathcal{S}$ is treated as the training set. A subset of $\mathcal{S}$, denoted as $P = \{p_1, p_2, ..., p_n\} \subseteq \mathcal{S}$, is chosen as the **prototype set**, where $n \leq N$. Prototypes serve as reference points in the embedding space. For a given input graph g, which can belong to $\mathcal{S}$ or any other set such as validation set or test set, compute the dissimilarity with each prototype graph $p_i \in P$. This results in n dissimilarities: $d_1 = d(g, p_1), d_2 = d(g, p_2), ..., d_n = d(g, p_n)$. Arrange these dissimilarities into n-dimensional vector: $d(g) = (d_1, d_2, ..., d_n)$.

Definition 7.7 ***(Graph Embedding)*** *For a graph domain $\mathcal{G}$, let $\mathcal{S} = \{g_1, g_2, ..., g_N\} \subseteq \mathcal{G}$ be a training set having N graphs. Let $P = \{p_1, p_2, ..., p_n\} \subseteq \mathcal{S}$ be a prototype set having $n \leq N$ graphs, then embedding is defined as the mapping*

$$\varphi_n^P : \mathcal{G} \to \mathbb{R}^n$$

where $\varphi_n^P(g) = (d(g, p_1), d(g, p_2), ..., d(g, p_n))$.

The resulting vector space has n axes, each corresponding to one prototype graph $p_i \in P$. The coordinate values for an embedded graph g represent its distances to these prototypes. Any graph from the training set $\mathcal{S}$ or any other set can be embedded into this vector space, making the embedding process highly flexible and extensible. While the GED is commonly used due to its robustness, other graph dissimilarity measures can also be applied. This allows the method to handle diverse graph types, including directed, undirected, labeled, or unlabeled graphs, as well as graphs with nodes and edges labeled from finite or infinite domains. The flexibility of the GED ensures resilience against various distortions in graph structure or labeling. By embedding graphs into a vector space, complex graph-based operations are reduced to simpler vector-based computations, enabling the use of standard machine learning algorithms. The embedding method can accommodate extensions to larger graph sets, making it suitable for real-world scenarios with dynamic or evolving datasets.

Dissimilarity embedding bear a strong resemblance to kernel methods. In kernel methods, patterns are characterized using pairwise kernel functions that quantify similarity. In contrast, the dissimilarity approach describes patterns through pairwise dissimilarities. However, there is a key distinction between the two methodologies. In kernel methods, kernel values are implicitly interpreted as dot products within a feature space. The pattern recognition task is performed within this implicit kernel feature space through kernel-based algorithms. On the other hand, the dissimilarity embedding approach explicitly constructs a vectorial representation of the pattern based on its dissimilarities to other patterns. This method does not rely on the assumption of an implicit feature space but instead creates a tangible and explicit dissimilarity space for analysis.

Embedding facilitate the use of vector-based classification methods. It enables efficient computation of graph similarities in the embedded space. It also simplifies the task of identifying patterns in graph data by leveraging vector-based representations and allow for clustering and visualization of graphs in a lower-dimensional space.

The dissimilarity-based graph embedding approach transforms graphs into vectors by calculating their distances to a set of prototype graphs. This transformation bridges the gap between graph-structured data and traditional vector-based machine learning techniques, providing a robust, flexible, and scalable method for analyzing and processing graph data. By leveraging the power of graph edit distance or similar measures, this approach is well-suited for diverse applications and capable of handling a wide variety of graph types.

Chapter 8

Graph Matching in Image Analysis

Graph-based methods are at the forefront of many image analysis tasks, thanks to their ability to represent complex structures and relationships. Graph matching, a technique for finding correspondences between the nodes or edges of two graphs, plays a crucial role in image analysis, where images are often represented as graphs to capture their structural properties. This chapter explores the concept of graph matching, its mathematical foundations, and its applications in image analysis. Image analysis is the process of examining and processing images to extract relevant information, patterns, or features. It is a multidisciplinary field that combines computer vision, pattern recognition, and machine learning techniques. Image analysis is widely used in various applications, including medical imaging, remote sensing, robotics, and multimedia.

An image is a representation of visual information. In the digital domain, an image is typically represented as a two-dimensional array of pixels, where each pixel has a value that represents a specific intensity or color. Images can be captured by cameras, scanners, or other imaging devices and are used in various forms, such as photographs, videos, and medical scans.

Image analysis is crucial for interpreting visual data in a wide range of applications. By analyzing images, computers can perform tasks that require visual understanding, such as identifying objects in a scene, detecting anomalies, or understanding spatial relationships. Image analysis also plays a critical role in automating processes that involve visual inspection, such as quality control in manufacturing or diagnosing medical conditions.

Image processing involves manipulating and enhancing images to improve their quality or extract specific features. Several techniques are used in image processing, each serving different purposes. Before analyzing an image, it is often necessary to preprocess it to enhance its quality and make it suitable for analysis. Common preprocessing steps are noise reduction, contrast enhancement, image smoothing and image sharpening. Noise in an image can obscure important features. Techniques such as Gaussian filtering or median filtering are used to reduce noise while preserving the edges of objects. Enhancing the contrast of an image makes it easier to distinguish between different regions. Histogram equalization is a widely used method for this purpose. Smoothing reduces sharp transitions in an image, which can help in reducing noise and making object boundaries more distinct. Gaussian blur is a common smoothing technique. In contrast to smoothing, sharpening enhances the edges in an

DOI: 10.1201/9781003532248-8

image, making the boundaries of objects more pronounced. Techniques like the Laplacian filter are used for sharpening.

Edges are important features in an image as they represent the boundaries between different regions. Edge detection algorithms identify these boundaries by detecting discontinuities in pixel intensity. Common edge detection methods include: Sobel operator, Canny edge detector, Prewitt and Roberts operators. ***Sobel operator*** uses convolution with a kernel to approximate the gradient of the image intensity, highlighting regions of high-spatial frequency that correspond to edges. The ***Canny edge detector*** is a multi-stage algorithm that detects a wide range of edges in images. It involves noise reduction, gradient calculation, non-maximum suppression, and edge tracking by hysteresis. Prewitt and Roberts Operators are simpler versions of the Sobel operator, which also detect edges by calculating the gradient of the image.

This chapter is organized as follows. Section 8.1, describes image representation using graphs. It introduces pixel level graphs, region level graphs and hierarchical graphs among others used for representation of images. Section 8.2, presents graph-based object recognition techniques. It explains the various steps involved in the process of object recognition from preprocessing to graph matching to post-processing. Graph-based image retrieval is described in section 8.3, which discusses text-based, content-based and semantic-based image retrieval system also. Finally, section 8.4 describes important application of graph matching in image analysis such as image registration, shape analysis, motion analysis, etc.

8.1 Image Representation using Graphs

Representing images using graphs is a powerful and flexible approach in computer vision and image analysis. A graph-based representation captures both the structural and relational information within an image, providing a framework for various image processing and pattern recognition tasks.

An image is represented as a graph $G = (V, E)$, where V represents set of vertices and E represents set of edges connecting the vertices. Vertices or nodes can represent image elements such as pixels, superpixels or regions. Each node can be associated with attributes or features, such as intensity, color, texture, or spatial coordinates. Edges represent relationships or interactions between nodes, such as spatial adjacency or similarity in feature values. Edges can be weighted to encode the strength of the relationship, such as distance or feature similarity.

Image representation can be broadly classified into four levels based on data organization. The Figure 8.1 illustrates a conceptual flow of information, progressing from signals with minimal abstraction to highly abstract descriptions necessary for image understanding. However, this process is not strictly

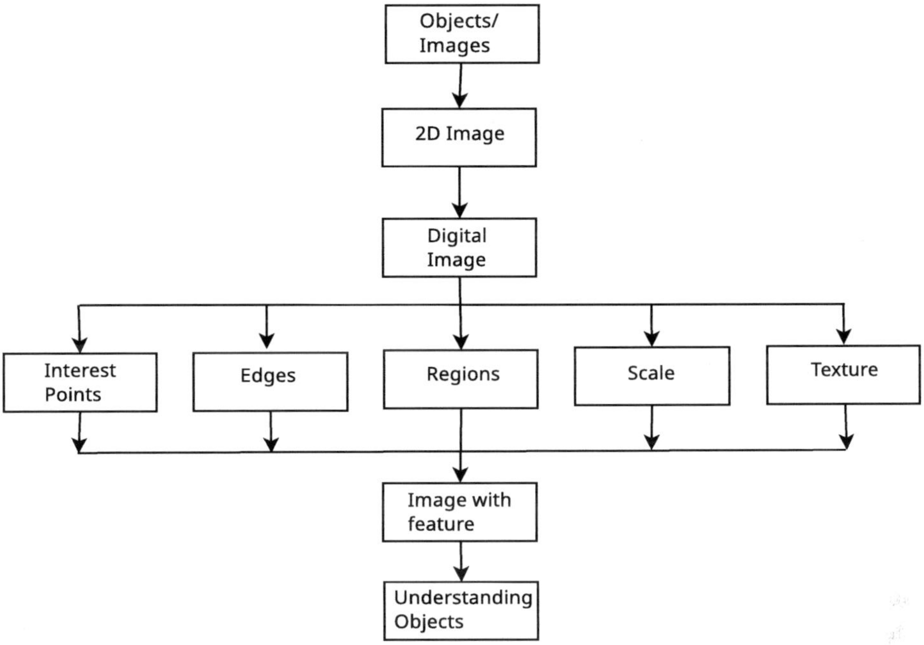

FIGURE 8.1: Levels of Image Representations in Image Analysis

unidirectional; feedback loops are often incorporated to allow algorithms to be adjusted based on intermediate results. This hierarchical structure of image representation and associated algorithms is often simplified further. Typically, two primary levels are distinguished: low-level image processing and high-level image understanding. Low-level image processing involves minimal knowledge of image content. When content-related information is available, it is usually derived from high-level algorithms or directly provided by a human expert familiar with the domain. Low-level techniques include tasks such as image compression, noise filtering, edge detection, and image sharpening. These methods process data that closely resemble the original input image. For instance, an image captured by a camera is inherently two-dimensional and is represented by an image function $f(x, y)$, where x and y denote spatial coordinates, and the function's value typically corresponds to brightness.

Before an image can be processed by a computer, it must first be digitized. Once digitized, it is commonly represented as a rectangular matrix, where each element corresponds to the brightness at a specific location in the image. In modern applications, this representation is often more complex, as images are typically in color, comprising three channels—red, green, and blue. Additionally, images frequently exist as part of a video sequence with an associated frame rate. Regardless, the raw data ultimately consists of a series of matrices that serve as the input and output of low-level image processing techniques.

In contrast, high-level image processing relies on knowledge, objectives, and strategies to achieve specific goals, often incorporating artificial intelligence techniques. High-level computer vision seeks to replicate human cognition and decision-making based on image content. For example, in recognizing an object such as a cow, high-level knowledge encompasses its shape and the complex relationships among its various parts, as well as their dynamic interactions. This process begins with a structured model of the world, against which digitized image data is compared. When discrepancies arise, the system attempts to resolve them by identifying partial matches or sub-goals, subsequently delegating specific tasks to low-level image processing.

This iterative process fosters a cooperative interaction between top-down (knowledge-driven) and bottom-up (data-driven) approaches, ultimately leading to image understanding. A feedback loop is established, wherein high-level results guide low-level processing tasks, and the system continues refining its interpretation until it converges on a comprehensive understanding of the image.

There are several types of graph representation for images. In ***pixel-level graphs*** each pixel in the image is represented as a node. Edges connect neighboring pixels. It is useful for low-level processing tasks such as segmentation and filtering. Figure 8.2 shows 4-neighborhood and 8-neighborhood of a representative gray filled pixel in the middle. ***Region-level graphs*** is another representation for images in which nodes correspond to larger regions in the image, such as superpixels or segmented regions. Edges capture relationships between regions, such as boundary adjacency or feature similarity. This representation reduces computational complexity compared to pixel-level graphs. Examples of region-based graph representations are region adjacency graph and superpixel graph. Region adjacency graph represents nodes as regions obtained via segmentation and edges reflects connecting adjacent regions. Edge weight denoted boundary strength or similarity between regions. In superpixel graph nodes represent superpixel, and edges denote connecting adjacent superpixels. Edge weight depicts color, texture or boundary similarity. This representation preserves image structure while reducing complexity. In ***Object-level graph representation*** model entire objects and their

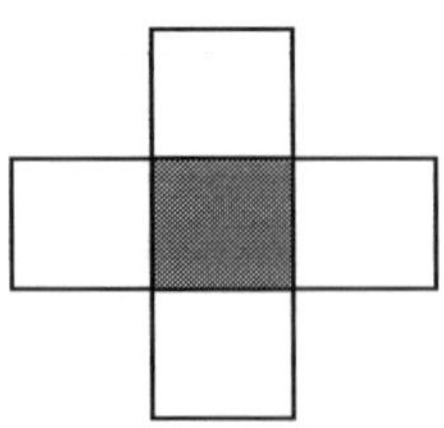

(a) 4-neighborhood

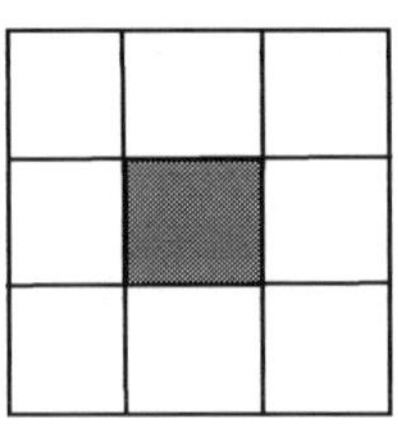

(b) 8-neighborhood

FIGURE 8.2: Neighborhood of a Pixel

relationship. Attributed graph is an example object-level graph representation. An Attributed Graph is a type of graph where nodes, edges, or both are associated with additional information in the form of attributes. These attributes can represent various properties or characteristics that provide richer information about the graph structure and its elements.

An ***Attributed graph*** is defined as:

$$G = (V, E, \mathcal{X}, \mathcal{Y})$$

where:

- V is the set of vertices
- $E \subseteq V \times V$ is the set of edges connecting vertices
- $\mathcal{X} : V \to \mathbb{R}^d$ assigns a d-dimensional attribute vector to each node
- $\mathcal{Y} : E \to \mathbb{R}^k$ assigns a k-dimensional attribute vector to each edge

Therefore, each node $v \in V$ has an associated feature vector $\mathcal{X}(v)$, and each edge $e \in E$ has an associated attribute $\mathcal{Y}(e)$. Attributed graphs can be categorized based on where attributes are assigned. In node-attributed graphs, attributes are associated with nodes only. Example of a node-attributed graph is social network where each user (node) has attributes like age, location, and interests. Edge-attributed graphs have attributes associated with edges only. For example a transportation network where edges (routes) have attributes like distance, speed limit, and traffic conditions. In fully attributed graphs, both nodes and edges have attributes. For example a molecular graph where nodes (atoms) have chemical properties and edges (bonds) have bond types or energy values.

In ***keypoint graphs*** nodes represent keypoints or interest points detected in the image while edges connect keypoints based on spatial proximity or feature similarity. It is commonly used in tasks like object recognition and matching. Another representation for image is ***hierarchical graphs*** which combine different levels of representation, such as pixels, regions, and objects, into a single hierarchical graph structure. This representation facilitate multi-scale analysis and high-level scene understanding.

Graph-based image representation capture structural relationship as graphs naturally encode spatial and topological relationships between image elements, which are critical for tasks like segmentation and recognition. Nodes and edges can store various attributes, enabling a rich representation of image properties. By adjusting the granularity of nodes (e.g., using regions instead of pixels), graphs can scale to represent large images efficiently. Unlike grid-based representations, graphs can represent irregular and non-uniform structures, such as objects with complex shapes.

The applications of graph-based image representation include image segmentation and object recognition. Other applications are image retrieval, scene understanding, image compression and medical imaging. For example, to construct graph from an image, the first step is to extract features such as intensity, color, texture, or keypoints. The next step is to define nodes and edges. Represent each pixel, superpixel, or feature point as a node. Connect nodes based on adjacency or similarity (e.g., using spatial distance or feature correlation). Then assign weights to edges based on the strength of the relationship (e.g., Euclidean distance, color similarity).

Graph-based image representation is a versatile and robust approach for capturing the relationships and structures inherent in images. By encoding spatial, topological, and feature-based information, graphs provide a comprehensive framework for addressing diverse image processing and analysis tasks. Their flexibility and scalability make them particularly suitable for complex problems in computer vision and machine learning.

8.1.1 Image Segmentation

Image segmentation is the process of dividing an image into distinct regions, each corresponding to a different object or part of an object. Segmentation is essential for tasks such as object recognition and image classification. Techniques for image segmentation include: thresholding, clustering and graph-based segmentation etc. Figure 8.3 shows the important techniques of image segmentation.

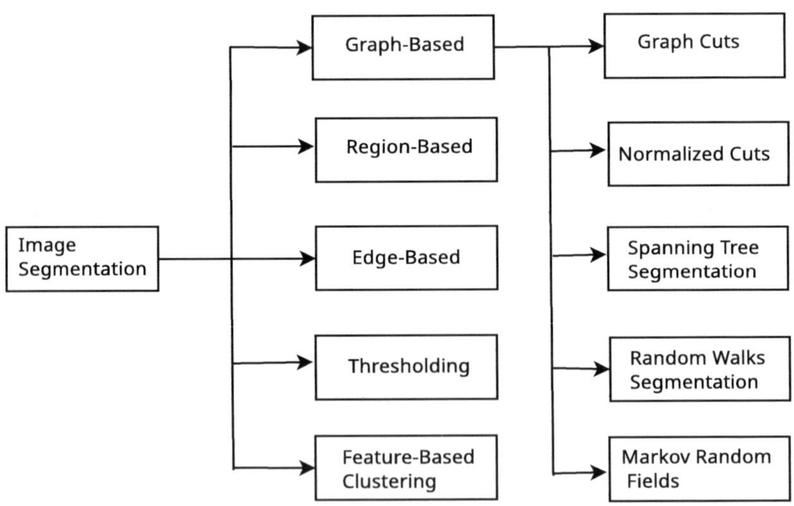

FIGURE 8.3: Image Segmentation Techniques

Thresholding is the simplest segmentation technique, where pixels are divided into regions based on their intensity values. Global thresholding uses a single threshold for the entire image, while adaptive thresholding changes the threshold value based on local image properties. ***Clustering*** algorithms like K-means can segment an image by grouping pixels with similar features (e.g., color or texture) into clusters, each representing a different segment. ***Edge-based segmentation*** techniques rely on detecting edges first and then using these edges to define the boundaries of segments. ***Region-based methods***, such as region growing and watershed algorithms, segment an image by starting with a seed point and growing the region by including neighboring pixels that have similar properties.

Graph-Based Segmentation

Graph-based segmentation is a technique that models an image as a graph, where pixels, superpixels, or regions are treated as nodes, and their relationships (e.g., spatial or feature similarities) are represented as edges. The goal is to partition the image into meaningful regions by leveraging graph structures and optimization algorithms. Important graph-based segmentation techniques are graph cuts, minimum spanning tree segmentation, normalized cuts, random walk segmentation and Markov random fields [68]. ***Graph cuts*** is a powerful image segmentation technique based on graph theory. It is widely used in computer vision and medical imaging to partition an image into meaningful regions. This method treats segmentation as a flow network problem. It formulates segmentation as a min-cut problem on a graph, where the goal is to find an optimal separation between foreground and background pixels by minimizing a cost function. In its flow network formulation, nodes of the graph corresponds to pixels or superpixels, while the edges represent connecting neighboring pixels. Each edge (p, q) has a weight that represents the similarity between pixels p and q. Two special terminal nodes are used, first node Source (S) represents the foreground, last node Sink (T) represents the background. Graph cuts use an energy function to model segmentation: $E(L) = R(L)+B(L)$, where $R(L)$ is regional term which assigns pixels to foreground/background based on likelihoods, and $B(L)$ is boundary term which ensures smoothness by penalizing discontinuities. Regional term determines the likelihood of a pixel belonging to a particular class using intensity, color histograms or Gaussian mixture models.

$$R(p) = -\log P(I_p|L_p)$$

where $P(I_p|L_p)$ is the probability of intensity I_p given the label L_p. Boundary term penalizes cutting through strong edges representing discontinuities.

$$B(p, q) = exp\left(-\frac{|I_p - I_q|^2}{2\sigma^2}\right)$$

where, I_p, I_q are pixel intensities and σ controls edge sensitivity. Segmentation problem is solved using the max-flow/min-cut algorithm such as Ford-Fulkerson algorithm using which the minimum cut partitions the graph into two disjoint sets. Pixels assigned to Source (S) corresponds to foreground whereas pixels assigned to Sink (T) represent background. The final segmentation mask is obtained from the min-cut result. The foreground region contains pixels connected to S, and the background contains pixels connected to T.

Minimum spanning tree (MST) segmentation technique is a graph-based image segmentation method that partitions an image into regions by constructing a MST over the graph representation of the image. It constructs an MST from the graph and iteratively removes the highest-weight edges to form regions. It is widely used in computer vision, medical imaging, and object detection due to its ability to preserve edges and structures while efficiently segmenting an image. Given an image represented as a graph $G = (V, E)$, the goal of MST-based segmentation is to partition the image into meaningful regions by constructing a MST and cutting edges based on a segmentation criterion. Here, vertices represents pixels or superpixels and edges connects two neighboring pixels. Weight function $w(p, q)$ measures the dissimilarity between adjacent pixels p and q. MST T is a subset of edges $E' \subseteq E$ that connects all pixels with the minimum total edge weight. Each pixel in the image corresponds to a node in the graph, and edges are formed between neighboring pixels (typically 4-connected or 8-connected neighborhoods). The edge weights $w(p, q)$ define the similarity between two pixels and are often computed using the following wight functions. *Intensity difference-based* weights is represented by

$$w(p, q) = |I(p) - I(q)|$$

where $I(p)$ and $I(q)$ are intensity values of pixels p and q. A larger difference corresponds to a stronger edge, indicating a possible boundary. *Color distance-based* weights denotes edges by

$$w(p, q) = ||C(p) - C(q)||_2$$

where $C(p)$ and $C(q)$ are color vectors in RGB of Lab space. *Texture-based* edge weight is denoted by

$$w(p, q) = ||T(p) - T(q)||_2$$

where $T(p)$ and $T(q)$ are texture feature vectors. Combined edge weight function is represented by

$$w(p, q) = \alpha|I(p) - I(q)| + \beta||C(p) - C(q)||_2 + \gamma||T(p) - T(q)||_2$$

where α, β, γ are weighting parameters.

To segment the image, we remove edges with large weights, which represent strong intensity, color, or texture changes. In *fixed thresholding approach*, we define a threshold τ, and removes edges with weight exceeding τ

$$E' = \{(p, q) \in T | w(p, q) \leq \tau\}$$

Adaptive thresholding approach is a graph-based segmentation algorithm that adapts the segmentation threshold dynamically

$$w(p,q) > MST_k(p,q) + \frac{k}{|C|}$$

where $MST_k(p,q)$ is the minimum weight edge in the MST for the segment containing p and q, $|C|$ is the size of the segment and k is a tuning parameter that controls segmentation granularity. *Normalized Cut Criterion* is an advanced method that minimizes the ratio of cut edges to the total edge weight:

$$N_{cut}(A,B) = \frac{\Sigma_{p\in A,q\in B} w(p,q)}{\Sigma_{p\in A,q\in A} w(p,q) + \Sigma_{p\in B,q\in B} w(p,q)}$$

where A and B are segments.

To compute the computational complexity analysis of MST segmentation method, we observe that graph construction takes $O(|E|)$ time. MST Kruskal or Prims will take $O(|E|\log|V|)$ time whereas segmentation will take $O(|V|)$ time. Therefore overall complexity becomes $O(|E|\log|V|)$, which is quiet efficient for large images.

Normalized cuts is a graph-based image segmentation technique used to partition an image into meaningful regions [174]. This technique formulates segmentation as a graph partitioning problem. It minimizes the dissimilarity between different groups while maximizing intra-group similarity. This method uses spectral clustering on the graph's Laplacian matrix. Unlike traditional clustering methods that focus on local features, normalized cuts consider the global structure of an image by treating it as a graph and partitioning it based on similarity measures. Given a graph $G = (V, E)$, the goal is to partition it into two disjoint sets A and B such that $V = A \cup B$ and $A \cap B = \phi$. The standard graph cut is defined as the sum of edge weights connecting two sets:

$$Cut(A,B) = \sum_{i\in A, j\in B} w(i,j)$$

where $w(i,j)$ represents the similarity weight between nodes i and j. This weight is typically computed using a similarity function based on color, texture, or spatial proximity. The above cut does not account for imbalanced segment sizes. Normalized Cuts (N-Cuts) introduce a normalization term to ensure a balanced partitioning:

$$N_{cut}(A,B) = \frac{Cut(A,B)}{Assoc(A,V)} + \frac{Cut(A,B)}{Assoc(B,V)}$$

where $Assoc(A,V)$ is the sum of all edge weights connecting nodes in A to the entire graph V: $Assoc(A,V) = \sum_{i\in A, j\in V} w(i,j)$. This normalization prevents the algorithm from producing small, meaningless segments by favoring balanced partitions.

Random walks segmentation technique is a graph-based image segmentation method that assigns each pixel to a specific segment based on the probability of a random walker reaching a predefined seed point. This method is particularly effective for interactive image segmentation, where the user provides a few labeled pixels (seeds), and the algorithm propagates the segmentation accordingly [85]. A random walker is a virtual agent that moves from one pixel (node) to another in an image, following weighted edges that represent pixel similarity. The key idea is that if a random walker starting at an unlabeled pixel is more likely to reach a seed of class A rather than class B, that pixel is assigned to class A. The probability of reaching a specific seed is computed using the graph Laplacian, ensuring smooth transitions between similar pixels. The Laplacian matrix is defined as:

$$L_{ij} = \begin{cases} d_i & \text{if } i = j \\ -w_{ij} & \text{if } v_i \text{ and } v_j \text{ are adjacent nodes} \\ 0 & \text{otherwise,} \end{cases}$$

where $d_i = \Sigma w(e_{ij})$ is the degree matrix and w_{ij} is weight matrix. Let s be the set of seeded pixels, and u be the unseeded pixels. The probability $p(i)$ of an unseeded pixel belonging to a certain label is obtained by solving the equation: $L_{uu}p_u = -L_{us}p_s$, where L_{uu} corresponds to unseeded pixels, L_{us} represents connections between unseeded and seeded pixels and p_s are the labels of seeded pixels. This is a system of linear equations that can be efficiently solved to obtain probability values for unseeded pixels. Each pixel is assigned to the class with the highest probability. Finally, output the segmented image with smooth boundaries.

Markov random fields is a probabilistic graphical model used for image segmentation and other vision tasks. It models images as undirected graphs, capturing spatial dependencies between pixels. By enforcing local smoothness constraints, Markov random field based segmentation methods ensure that neighboring pixels with similar properties belong to the same segment. This method is often combined with deep learning for fine-grained segmentation. Markov random field is defined over a set of random variables where each variable is conditionally dependent on its neighbors. The segmentation problem is formulated as energy minimization using the Gibbs distribution:

$$P(L) = \frac{1}{Z} e^{-E(L)}$$

where $P(L)$ is the probability of a labeling L, $E(L)$ is the energy function that needs to be minimized, Z is a normalization constant (partition function). The energy function itself consists of: $E(L) = E_{data}(L) + \lambda E_{smoothness}(L)$, where $E_{data}(L)$ measures how well the assigned labels match observed pixel intensities or features, and $E_{smoothness}(L)$ ensures neighboring pixels have similar labels to prevent noise, λ is a weighting factor balancing data fidelity and smoothness.

Graph-based image segmentation offers many advantages. It captures global and local relationships by modeling pixel dependencies efficiently. Graph-based image segmentation is flexible and adaptive which works well with different feature types such as color, texture or shape. It can also handles irregular structures and better suited for complex segmentation tasks. Applications of graph-based image segmentation are in object detection, scene understanding and medical image analysis, etc.

8.2 Object Recognition

Object recognition using graph matching is a technique that models objects and their structural relationships as graphs and identifies objects by comparing these graphs. Figure 8.4 describe the steps in processing of an object recognition system. The objective of object recognition is to identify and locate all instances of a specific set of objects within an image. These objects may belong to various categories, each characterized by unique shapes. Typically, their number and positions within the image are not predetermined. Complicating this task, images often contain intricate background elements that must be disregarded. In certain cases, objects of interest may be partially obscured by other objects or by the background itself.

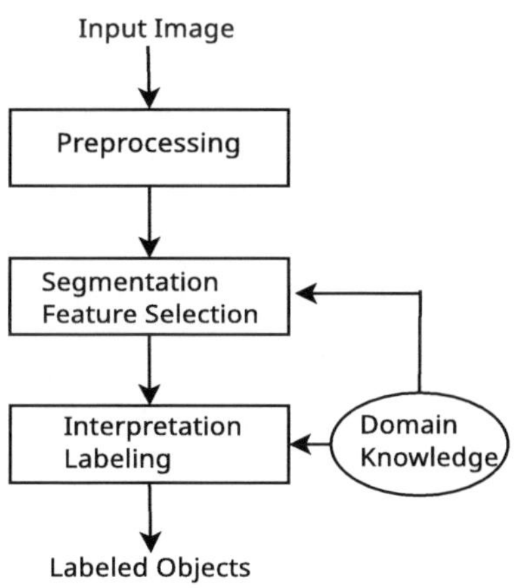

FIGURE 8.4: Object Recognition System

Graph-based object recognition addresses these challenges by breaking down the image into smaller, manageable components and representing these as a graph. This graph encodes the relationships between components. The recognition process involves searching for subgraphs within the image graph that correspond to the target object's graph representation, employing inexact graph matching techniques to handle variations and distortions. Graph matching provides a robust framework for recognizing objects, particularly in situations where the objects exhibit variations in appearance, scale, or orientation.

Object recognition using graph matching involve a sequence of steps: graph construction, feature extraction, graph matching, decision making and post processing. During ***graph construction*** the input image is represented and modeled as a graph. This step defines nodes and edges, and assign attributes.

Feature extraction process detect key points or segment regions of interest. This step extract features, assign descriptive features to nodes (e.g., color, texture) and geometric or relational attributes to edges. Various types of features are employed in object recognition, primarily categorized as region-based or boundary-based. It is generally assumed that a region or a closed boundary corresponds to an entity, either an entire object or a part of one. The commonly used features in object recognition can be classified as local features, global features and relational features. ***Local features*** are typically associated with an object's boundary or a distinct small area within a region. Properties such as curvature and its derivatives are frequently used as local features. High-curvature points, often referred to as corners, play a crucial role in object recognition. Local features may also describe the shape of small boundary segments or surface patches, with common examples including curvature, boundary segments, and corner points. ***Global features*** characterize entire regions within an image, including attributes such as area (size), perimeter, Fourier descriptors, and moments. These features can be extracted either by analyzing all points within a region or by focusing solely on the boundary points. In both cases, the objective is to derive descriptors that consider all points, their spatial distribution, intensity variations, and interrelations. ***Relational features*** capture the relative positioning of different elements, such as regions, closed contours, or local features. These features often include measurements of distances between features and relative orientations. They are particularly valuable for defining composite objects composed of multiple regions or local features within an image. In many cases, the spatial arrangement of features is what determines an object's identity. The same set of features, when arranged differently, can represent entirely distinct objects.

Graph matching Align the input graph with the model graph using an appropriate graph matching algorithm (e.g., graph edit distance, spectral matching). ***Decision making*** step determine whether the model graph exists in the input graph based on similarity scores or matching costs. It also incorporate thresholds or probabilistic measures to handle ambiguities. ***Post-processing*** step refine results to handle errors due to noise, occlusion, or partial visibility. Figure 8.5 illustrates the steps of the graph-based object recognition system.

A fundamental approach to graph-based object recognition uses a low-level representation where each pixel in the image is treated as a graph node. The labels for these nodes are determined using the topographical primal sketch (TPS). A TPS assigns each pixel a label that remains invariant under

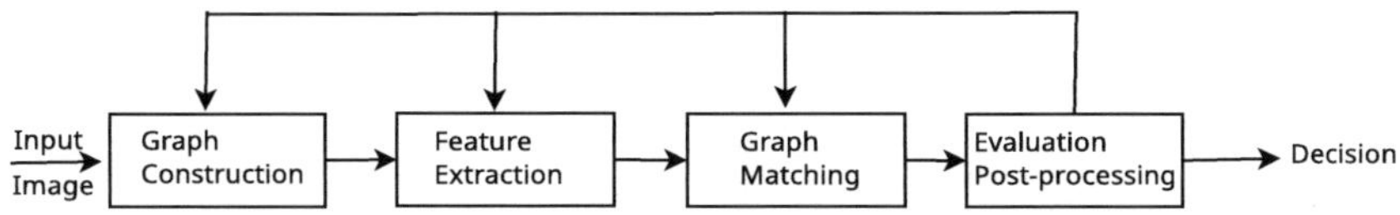

FIGURE 8.5: Graph-based Object Recognition

monotonic transformations of the image's gray levels. This is achieved by fitting a local two-dimensional cubic surface to the intensity values around each pixel, which estimates the intensity surface.

The graph matching techniques proposed for object recognition in this framework include distance-based matching and feature-based matching. ***Distance-based matching*** computes a distance measure between node labels in the test and model graph. Both the test image and the model image are first aligned (registered) based on the positions of node labels. A cost is calculated based on the relative distances between nodes that share the same label in the test and model images. ***Feature-based matching*** is another approach, where a feature vector is associated with nodes that have specific labels. The feature vector is derived by calculating the second derivative extrema, the directions of these extrema, and the gradient around the node. A similarity measure is then computed based on these feature vectors to compare the test and model graphs.

A novel mid-level shape descriptor called shape context for measuring the shape similarity is introduced in [10]. The method begins by extracting the edges from an image and selecting N uniformly spaced points along these edges. For each sampled point, a compact descriptor is created by computing a coarse histogram of the relative coordinates of the remaining points. This histogram is defined in a log-polar coordinate system, which capture both angular and radial distribution of the points relative to the sampled point. The resulting histogram for all N points are flattened and concatenated to form the shape context representation of the image. In addition to this global shape representation, a complementary descriptor based on local appearance is used. Specifically, the tangent angle at each of the N points is calculated to provide local orientation information. When these points are represented as nodes in a graph, the cost of matching two nodes can be quantified using the dissimilarity between their respective shape context histograms and the difference in their tangent angles. The object recognition task is then framed as a weighted bipartite graph matching problem, where the goal is to find the optimal correspondence between points in two images. This problem is efficiently solved using the Hungarian method, which minimizes the overall matching cost. This approach effectively combines global shape information and local appearance features, making it a robust method for shape-based object recognition.

Li and Lee proposed a method in [121] where a graph represents a 2D scene described through a polygonal approximation. In this presentation nodes correspond to the vertices of the polygon and edges represent the polygon's sides. Node attributes are defined by the angles subtended at each vertex. Edge attributes are determined by the distance between connected nodes. This graph is referred to as the ***scene graph*** and is used to model the 2D scene. To handle distortions and occlusions effectively the scene graph is divided into smaller component called *sub-scene* graphs. Each sub-scene graph captures a localized portion of the scene. The recognition process involves performing inexact sub-scene graph matching between each sub-scene graph and a predefined ***model graph***. The matching is accomplished using a Hopfield neural network, which is capable of handling variations and imperfections in the data. The overall matching for the complete scene graph is then determined by aggregating and analyzing the statistical results of matching all the sub-scene graphs to the model graph. This approach ensures robustness in recognizing complex scenes, even in the presence of distortions or partial occlusions.

Sebastian *et al.* [167] proposed a system for recognizing object shapes based on their silhouettes. Their method represents each object as a ***shock tree***, a structure derived from the thinning of the shape. The shock tree captures the hierarchical organization and geometric properties of the silhouette. For matching, they introduce a tree edit distance metric, where the edit costs are not arbitrarily assigned but are analytically derived from a small set of hypotheses. These hypotheses relate to the costs associated with deforming a silhouette, providing a principled basis for the distance calculation. The computation of this tree edit distance is performed using an error-correcting tree isomorphism algorithm. This algorithm, which leverages dynamic programming, efficiently handles structural variations and deformations, enabling robust matching between shock trees. This approach allows for accurate recognition of object shapes despite potential distortions in their silhouettes.

The problem of visual inspection using a high-level representation scheme is addressed by [112]. Their approach focuses on Printed Circuit Board (PCB) images, which are represented as trees. The process begins with ***binarization*** of the image, followed by partitioning the binarized image into non-overlapping regions, or blobs, each composed of adjacent pixels with the same value. A node is assigned to each blob, and a tree structure is constructed by adding edges between nodes if the corresponding blobs are spatially nested and have different pixel values. The root node of the tree represents the blob that encloses the outer boundary pixels of the image. To identify defects, the tree derived from a test image is compared with the tree generated from a defect-free reference image using a tree isomorphism algorithm. If the two trees are not isomorphic, a defect is detected, and the inspection process terminates. If the trees are isomorphic, a second inspection step is performed. This step involves extracting polygonal-boundary information for further analysis. A tolerance zone is defined, and the algorithm uses a polygonal-boundary matching function to verify if the pairs of matched nodes satisfy the tolerance criteria. Any deviation beyond the tolerance indicates a potential defect, ensuring a thorough inspection process.

Object recognition using graph matching is a versatile and powerful method that leverages the structural and relational information of objects. By representing objects as graphs and employing advanced matching algorithms, this technique excels in scenarios involving variations or complex patterns. Its ability to integrate both structural and feature-based information makes it indispensable for modern computer vision applications.

8.2.1 Graph Neural Networks

Graph Neural Networks (GNNs) are a class of deep learning models designed to operate on graph-structured data. Unlike traditional Convolutional Neural Networks (CNNs) that work on grid-like structures (e.g., images), GNNs generalize to non-Euclidean domains such as social networks, molecules, and object recognition tasks where objects and their relationships form a graph. Figure 8.6 shows structural representation of texts, images and networks. In object recognition, GNNs are used to capture relationships between objects, such as spatial arrangements, co-occurrence, and contextual dependencies, making them highly effective for tasks where structural information plays a critical role.

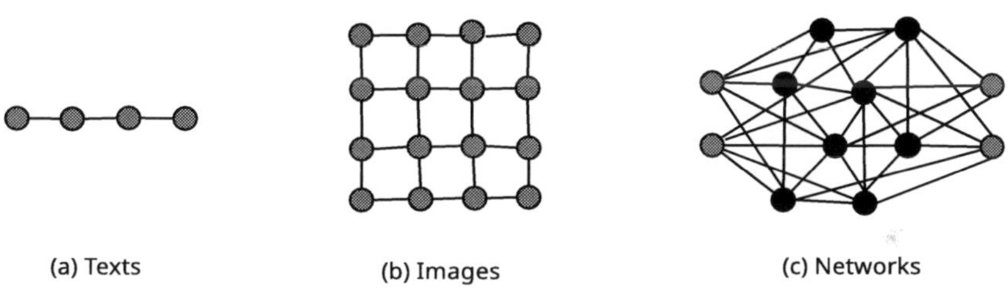

FIGURE 8.6: Structure of Texts, Images and Networks

Traditional object recognition methods rely on pixel-level features extracted from CNNs. However, they may struggle with complex object relationship like scenes with multiple interacting objects. GNNs preserve spatial and semantic relationships between object parts. GNNs handles irregular and non-Euclidean data. Unlike CNNs, GNNs can work with data that does not fit into a regular grid such as 3D point clouds and skeleton-based human pose. Since graphs encode relationships, GNNs can recognize objects with limited labeled data.

GNNs process graph-structured data by iteratively aggregating information from node neighbors. The core idea is to update a node's feature representation based on its neighbors. Each GNN layer updates node features using message passing. In GNNs ***message passing*** between nodes is defined as:

$$h_v^{(l+1)} = \sigma\left(W \cdot \sum_{u\in\mathcal{N}(v)} f(h_u^{(l)}, h_v^{(l)}, e_{uv})\right)$$

where:

- $h_v^{(l)}$ is the feature of node v at layer l.
- $\mathcal{N}(v)$ is the set of neighboring nodes.
- $f(\cdot)$ is an aggregation function (e.g., sum, mean, max).
- W is a learnable weight matrix.
- σ is a non-linear activation function.

Message passing in GNNs depicts the sharing of information between nodes in graphs along the connecting edges.

Graph Convolutional Networks

Graph Convolutional Networks (GCNs) generalize CNNs to graph data using message passing between nodes.

$$h_v^{(l+1)} = \sigma\left(\sum_{u\in\mathcal{N}(v)} \frac{1}{c_{vu}} W^{(l)} h_u^{(l)}\right)$$

where:

- $h_v^{(l+1)}$ is the updated feature of node v at layer $l+1$.
- $W^{(l)}$ is the weight matrix.
- c_{vu} is a normalization factor.
- $\mathcal{N}(v)$ denotes the neighbors of v.

GCNs aggregates neighboring node features using a normalized adjacency matrix. GCNs can be applied in scene understanding by modeling object-object interactions.

Graph Attention Networks

Graph Attention Networks (GAT) improve GCNs by using self-attention mechanisms to weigh node importance dynamically:

$$h_v^{(l+1)} = \sigma\left(\sum_{u\in\mathcal{N}(v)} \alpha_{vu} W^{(l)} h_u^{(l)}\right)$$

where α_{vu} is the attention score between nodes v and u. GAT can be used for recognizing objects in complex scenes by prioritizing important relationships.

GNNs can be applied in object recognition using scene graphs, skeleton graphs and point clouds. *Scene graphs* represent objects as nodes and their relationships as edges. In scene graph-based object recognition, objects and their semantic relationships are modeled as a graph. GNNs refine object labels based on relational reasoning. *Skeleton Graphs*, represent human pose keypoints for action recognition. In skeleton-based object recognition humans skeletons are represented as keypoint graphs. GNNs analyze movement patterns for action classification. *Point clouds* are graph-based representations of 3D objects. In point clouds-based object recognition objects in 3D space are modeled as graphs for shape recognition. GNNs capture global and local geometric structures. Point clouds can be applied in 3D scene understanding robotics and autonomous vehicles. GNNs help recognize new objects with limited examples by propagating information across graph nodes.

GNNs are transforming object recognition by incorporating relational and structural information. Whether applied to scene graphs, human skeletons, or 3D object recognition, GNNs provide a powerful alternative to traditional deep learning models, especially in complex and structured environments.

Some recent works on GNNs applied to object recognition is described in [11, 87, 202]. Zhong *et al.* in [202] proposed an adaptive graph reasoning network for object detection. They construct a dynamic relation graph which makes a better integration of semantics and visual features using attention mechanism. Bera *et al.* in [11] introduce SR-GNN, which effectively captures subtle changes by aggregating context-aware features from relevant image regions. This approach enhances the differentiability of features in fine-grained visual classification tasks without relying on bounding-box or distinguishable part annotations. Han *et al.* in [87] propose to represent the image as a graph structure and introduces a Vision GNN (ViG) architecture to extract graph-level features for visual tasks. The authors build both isotropic and pyramid architectures of ViG and demonstrate its superiority in image recognition and object detection tasks. Authors in [106, 196] provides recent survey and overview of GNNs, including variants like graph recurrent networks, graph attention networks, graph convolutional networks and mutli-view learning.

8.2.2 Traditional Classification Techniques

Object recognition is a core task in image analysis and computer vision that involves identifying and classifying objects in images or videos. Classification techniques play a crucial role in this process by categorizing objects into predefined classes based on extracted features. Classification fundamentally involves recognizing objects based on their features. Pattern recognition methods belong to this category and have demonstrated their effectiveness across various applications. Classification techniques assume that features have been extracted from images and normalized to ensure their representation within

TABLE 8.1: Summary of Classification Techniques

Technique	Mathematical Concept
k-NN	Distance-based classification
SVM	Maximum margin hyperplane
Decision Trees	Information gain
Naïve Bayes	Bayes' Theorem
CNNs	Convolution
RNNs	Sequential processing

the same metric space. These techniques can be broadly categorized into traditional machine learning methods and deep learning-based methods. Figure 8.1 lists the summary of the important classification techniques.

Machine Learning-Based Classification Techniques

Traditional approaches to object recognition involve feature extraction followed by classification using conventional machine learning algorithms. ***k*-Nearest Neighbors** (k-NN) is a distance-based classifier that assigns a new object to the majority class among its k nearest neighbors in the feature space. k-NN is a non-parametric classification algorithm that assigns a class to a test sample based on the majority vote among its k nearest neighbors in the feature space. Given a dataset of labeled samples:

$$D = \{(x_1, y_1), (x_2, y_2), ..., (x_N, y_N)\}, \quad y_i \in \{C_1, C_2, ..., C_m\}$$

where x_i is the feature vector of the i-th sample and y_i is the corresponding class label.

For a test sample x^*, compute the Euclidean distance to each training sample:

$$d(x^*, x_i) = \sqrt{\sum_{j=1}^{d}(x_j^* - x_{ij})^2}$$

The class of x^* is determined by the majority vote among the k nearest neighbors.

Advantage of k-NN is that it is simple and intuitive. It also works well with small datasets; however it can be computationally expensive for large datasets. Performance of k-NN depends on the choice of k and distance metric such as Euclidean or Manhattan distance.

Naïve Bayes Classifier

Naïve Bayes Classifier is based on Bayes' theorem under the assumption that features are independent given the class label.

$$P(y|x) = \frac{P(x|y)P(y)}{P(x)}$$

Using the assumption of feature independence, the likelihood is expressed as:

$$P(x|y) = \prod_{j=1}^{d} P(x_j|y)$$

The decision rule is:

$$\hat{y} = \arg\max_{y} P(y) \prod_{j=1}^{d} P(x_j|y)$$

Naïve Bayes classifier is fast and efficient for high-dimensional data. Another example is that it works well for text and document classification.

Decision Trees and Random Forests

Decision trees are tree-like model that recursively splits features based on information gain or entropy to classify objects. Random forests are an ensemble of multiple decision trees that improves accuracy and robustness.
Entropy:

$$H(S) = -\sum_{i=1}^{m} p_i \log_2 p_i$$

where p_i is the proportion of samples belonging to class i.
Gini Impurity:

$$G(S) = 1 - \sum_{i=1}^{m} p_i^2$$

A ***random forest*** aggregates predictions from multiple decision trees:

$$\hat{y} = \frac{1}{T} \sum_{t=1}^{T} f_t(x)$$

where $f_t(x)$ is the prediction of tree t.

Decision tree model is easy to interpret and implement. It can handles both numerical and categorical data. On the other hand decision tree can sometime overfit small datasets.

Support Vector Machine (SVM)

SVM finds an optimal hyperplane that maximizes the margin between different object classes in a high-dimensional space. It works well with small and high-dimensional datasets. SVM is effective for both linear and non-linear classification (using kernel trick). Given training samples:

$$D = \{(x_1, y_1), (x_2, y_2), ..., (x_N, y_N)\}, \quad y_i \in \{-1, 1\}$$

the decision function of a linear SVM is:

$$f(x) = w^T x + b$$

The optimization problem is formulated as:

$$\min_{w,b} \frac{1}{2}||w||^2$$

subject to the constraint:

$$y_i(w^T x_i + b) \geq 1, \quad \forall i$$

For non-linearly separable data, the kernel trick maps data into a higher-dimensional space using a function $\phi(x)$, e.g., Radial Basis Function (RBF) kernel:

$$K(x_i, x_j) = e^{-\gamma||x_i - x_j||^2}$$

Deep Learning-Based Classification Techniques

Deep learning has revolutionized object recognition by enabling automatic feature extraction and classification in a single framework. ***Convolutional Neural Networks (CNNs)*** are hierarchical neural networks designed for processing spatial data (e.g., images). They use convolutional layers to extract features and fully connected layers for classification. CNN is a specialized deep learning architecture designed for processing structured grid data, such as images. It uses convolutional layers to automatically extract spatial features, reducing the need for manual feature engineering. CNNs are particularly effective in image recognition, object detection, and computer vision tasks.

CNNs applies a set of learnable filters (kernels) to input data. It computes feature maps by sliding the kernel over the image and computing element-wise multiplications and summations. For example a convolutional layer applies a kernel filter W to an input feature map X:

$$Y(i, j) = \sum_m \sum_n W(m, n) X(i - m, j - n)$$

A ReLU activation function introduces non-linearity:

$$f(x) = \max(0, x)$$

A fully connected layer applies:

$$y = \sigma(Wx + b)$$

TABLE 8.2: Comparison of Classification Techniques

Technique	**Feature Extr.**	**Performance**	**Complexity**
k-NN	Manual	Low	High
SVM	Manual	High	High
Decision Trees	Manual	Medium	Low
Random Forests	Manual	High	Medium
Naïve Bayes	Manual	Medium	Low
CNN	Automatic	Very High	High
RNN	Automatic	Medium-High	High

where σ is the softmax function for classification:

$$P(y_i|x) = \frac{e^{z_i}}{\sum_j e^{z_j}}$$

Recurrent Neural Networks (RNNs) and Long Short-Term Memory (LSTM) are used for sequential object recognition, such as gesture recognition in videos.

For an input sequence x_t, the hidden state at time t is:

$$h_t = \tanh(W_h h_{t-1} + W_x x_t + b)$$

LSTMs introduce memory cells with gates:

$$f_t = \sigma(W_f x_t + U_f h_{t-1} + b_f)$$

$$i_t = \sigma(W_i x_t + U_i h_{t-1} + b_i)$$

$$o_t = \sigma(W_o x_t + U_o h_{t-1} + b_o)$$

The cell state update is:

$$c_t = f_t \odot c_{t-1} + i_t \odot \tanh(W_c x_t + U_c h_{t-1} + b_c)$$

Table 8.2 shows the comparison of various classification techniques with feature extraction, performance, and computational cost.

The choice of classification technique for object recognition depends on the complexity of the dataset, the need for automatic feature extraction, and computational constraints. Traditional machine learning methods work well for small datasets, whereas deep learning techniques (CNNs, RNNs) dominate large-scale image recognition tasks. Emerging hybrid models continue to push the boundaries of object recognition performance.

8.3 Graph-based Image Retrieval

Information retrieval involves the acquisition, organization, storage, retrieval, and ranking of information. This task applies to various types of data, including text, images, videos, and music. An image is a visual representation of an object captured or rendered using different mediums or supports. ***Image Retrieval*** focuses on finding images from a collection based on user queries, with the goal of ranking similar images according to their relevance. This field has gained significant attention in domains like digital libraries, remote sensing, and astronomy, becoming highly active since the 1970s. Image Retrieval systems can be categorized into three main techniques: text-based image retrieval, content-based image retrieval and semantic-based image retrieval.

Text-Based Image Retrieval (TBIR) systems retrieve images using textual queries. This approach relies on indexing process, where textual descriptors, metadata or annotations are attached to images. Despite being foundational, research in TBIR is considered older due to growing focus on other image retrieval methods. TBIR uses textual indexing for tagging images with descriptive annotations. It often apply metadata or textual elements embedded with the image. This method lacks precision in capturing the visual aspects of images, leading to less effective retrieval in various applications. Also, the descriptive annotations often require manual input, making the process complex and labor-intensive.

Content-Based Image Retrieval (CBIR) has gained significant attention over the last few decades due to advancements in image acquisition devices, increased storage capacities, and high-quality digitization techniques. CBIR systems retrieve images based on visual features like color, texture, shape, and other characteristics. CBIR extracts visual descriptors from images and retrieves results based on visual similarity. It overcomes the limitations of TBIR by focusing on the image's content rather than textual annotations. It also allows the use of weighting functions to prioritize features based on system objectives and user needs. CBIR is useful in various fields, including medical imaging, remote sensing, and digital archives.

Semantic-Based Image Retrieval (SBIR) system performs image retrieval by associating images with semantic meaning. They aim to interpret the significance conveyed in an image, making it possible to retrieve images based on high-level concepts. SBIR extracts visual descriptors to identify significant regions within an image. It uses these descriptors to derive a semantic description, often involving knowledge extraction processes. It also bridges the gap between low-level features (e.g., color or texture) and high-level understanding (e.g., identifying objects or contexts). SBIR is useful in areas requiring contextual understanding, such as autonomous systems and advanced search engines. It is particularly effective in scenarios where visual and semantic cues combine to form meaningful retrieval criteria.

For several years, extensive effort has been devoted to studying image representation models, with the most commonly used approaches being vectors, strings, trees, and graphs. Among these Vectors are widely utilized in image retrieval. For instance, works such as Yang *et al.* [197] and Allani *et al.* [5] model images as vectors for retrieval tasks. Strings represent an ordered sequence of elements, often used in image retrieval where the ordering of elements is important. The distance between two strings is typically measured using the edit distance, as defined by Levenshtein [119] and applied in studies like Jenni *et al.* [103]. However, both vector-based and string-based approaches face limitations due to their inability to effectively model complex objects, making them unsuitable for certain applications. Tree-based models provide hierarchical representation, yet they are also impractical for capturing complex relationships between components. To address these limitations, many researchers have turned to graph-based models, which offers a more flexible and comprehensive framework for modeling images.

Graphs allow the representation of all possible relationships between components in an image. Unlike trees and strings, which are particular types of graphs, graph structures are not restricted to hierarchical or sequential relationships. The semantics associated with graph arcs are versatile, representing more than just typing or membership relations. A graph-based image is represented as a set of components (nodes) and a set of binary relationships between these components (edges). Figure 8.7 shows schematic diagram of graph-based image retrieval. Graph-based models are widely used in many applications due to their expressiveness in capturing both structure and semantics. Their advantages have made them a focal point of research for enhancing the efficiency and accuracy of IR systems. For instance, graph-based measures enable semantic representation of queries and documents and their application in semantic search models (e.g., [19], [128]).

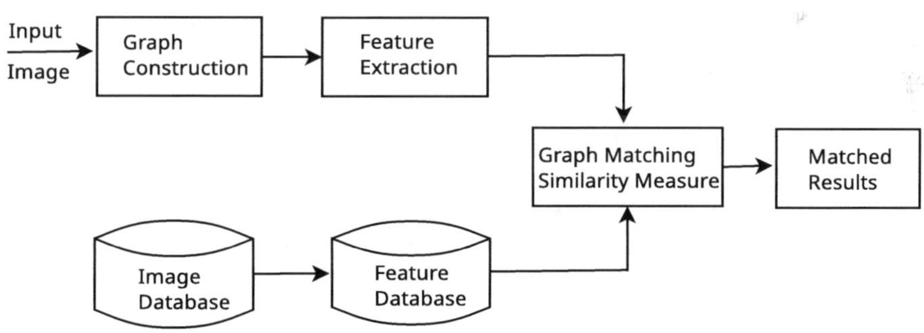

FIGURE 8.7: Graph-based Image Retrieval

Relevance feedback and query annotation are two important techniques that enhance the precision and effectiveness of graph-based image retrieval. ***Relevance Feedback*** technique improve retrieval performance by incorporating user-provided feedback on relevant or irrelevant images. For example,

Li and Hsu [120] designed an automatic relevance feedback system that classifies database images as relevant if they belong to the same semantic category as the query. The system demonstrated improved retrieval performance for images with clear regional correspondence. Rui *et al.* [158] highlighted how relevance feedback enables systems to refine semantic concepts and progressively enhance retrieval precision. ***Query Annotation*** technique involves introducing semantic annotations to influence the graph structure before retrieval. For example, Chen *et al.* [31] presented an annotation-based system using RDF triples to link concepts and instances. Semantic links between concepts address challenges in retrieval, enabling more accurate search results.

While traditional methods often focus on individual data features, they may overlook the underlying structural information in graphs, which is crucial for semantic discovery—especially in databases with cluster or manifold structures (Bin *et al.*, [16]). The semantics of a node in a graph can be influenced by its context, which includes ascendant and descendant nodes, requiring a broader interpretation of the image. Specific graph types used in graph-based image retrieval include: semantic graphs, conceptual graphs and scene graphs. Figure 8.8 illustrates a schematic diagram of semantic, conceptual and scene graphs. ***Semantic Graphs*** facilitates knowledge representation and reasoning. For instance, Shieh *et al.* [175] developed CKSGIS, a system that retrieves semantic graphs of related terms to assist users in finding images beyond a specific search term. ***Conceptual Graphs*** employed in semantic representation for information retrieval, as demonstrated in works like Gonçalves *et al.* [83] and Mechkour *et al.* [132]. A ***Scene Graphs*** is a structured representation of an image that describes the relationships between objects in a semantic and spatial manner. It provides a graph-based abstraction where nodes represent objects and edges capture relationships between them. An example of scene graph representation is provided in Figure 8.8. Scene graph represents the spatial structure of scenes logically, as shown in Johnson *et al.* [104].

For an image of a person sitting on a chair, with a cup on a table, the scene graph might consists of nodes, edges and attributes. Nodes consists of Person, chair, cup and table. Edges denotes the person sitting on the chair and cup on the table. Attributes may consists of Person wearing hat, or cup is white.

Graph-based image retrieval is an advanced technique that represents images as graphs and leverages graph analysis to find images similar to a query image. This method effectively captures both the visual and structural relationships within an image, enabling a robust approach to image comparison and retrieval. In graph-based retrieval, a query image is first represented as a graph. The retrieval system then searches for graphs in the database that are similar to the query graph.

Graph-based image retrieval involve several stages starting with ***graph construction*** which itself includes feature extraction and graph representation. ***Feature extraction*** detect features like keypoints or superpixels. It extract attributes for nodes (e.g., color, texture, shape) and edges (e.g., spatial

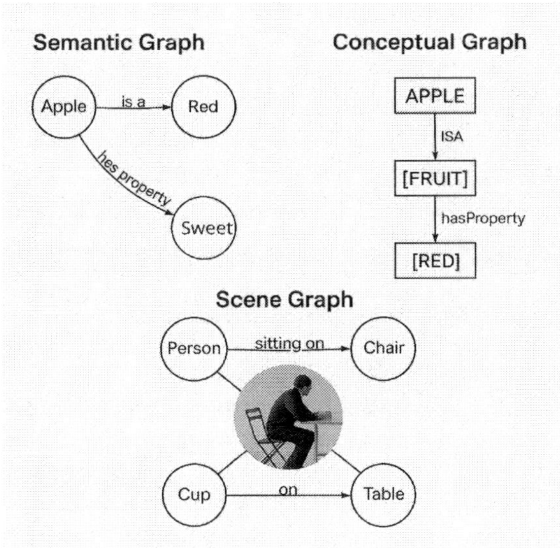

FIGURE 8.8: Semantic, Conceptual, and Scene Graphs

relationships, feature similarity). Graph representation construct the graph for each image based on extracted features and their relationship. In ***database preparation*** step each image is represented in the database as a graph and store graph representation. It optionally, preprocess the graphs for efficient matching, such as creating embedding or indexing them. During the ***query processing*** the query image is represented as a graph using the same process as the database images. Then the query graph is compared with the database graphs using graph matching techniques to measure similarity. Database images are *ranked* based on their similarity to the query graph. Finally, *retrieve* the top-ranked images as the results.

Graph-based image retrieval enables rich representation which captures both visual features and structural relationship within an image. It handles variations in scale, orientation, and viewpoint by focusing on relational patterns. Graph-based image retrieval allows flexibility by supporting the various type of images, including natural scenes, medical images, and diagrams. It also leads to distinguishing images more effectively by considering structural and contextual information.

Graph-based image retrieval is a powerful technique that leverages the structural and relational information encoded in images. By representing images as graphs and employing graph matching for similarity measurement, this approach provides a robust and flexible framework for retrieving visually and structurally similar images. Despite its computational challenges, advancements in graph embedding, indexing, and efficient matching algorithms continue to enhance its applicability in various domains.

8.4 Applications

Application of graph-based image retrieval includes medical imaging which requires retrieval of medical scans with similar anatomical structures for diagnosis and treatment planning. It is also applicable in document image retrieval in searching for images of documents, diagram, or maps with similar layout. Some other applications are scene understanding, trademark and logo retrieval. Scene understanding involves finding similar scenes or objects in surveillance and geographic imaging. It is also used in retrieving logos or trademarks based on structural pattern.

Graph matching is widely used in ***object recognition*** for recognizing objects in images by matching their graph-based representations to stored models. Objects in an image are represented as graphs where nodes correspond to keypoints, and edges capture spatial relationships or geometric features. For example it can be used for matching a scene graph to a database of model graphs to identify objects. It can also be used to recognize partially occluded objects through inexact graph matching. Applications of object recognition includes industrial inspection, autonomous navigation, and medical imaging.

Image registration involves aligning two or more images taken from different viewpoints, times, or sensors. Graphs are used to represent feature points in the images, and graph matching identifies correspondences between these points to align the images. As an example, it can be used in aligning satellite images for change detection. It can also be used in combining medical images from different modalities (e.g., MRI and CT). Other applications are remote sensing, medical diagnostics, and surveillance.

In ***shape analysis*** graph matching supports the analysis and comparison of shapes by capturing their structural and geometric properties. Shapes are represented as graphs, where nodes correspond to contour points or critical features, and edges encode geometric relationships such as angles or distances. For example analyzing biological shapes (e.g., cells or organs) for classification and comparing product designs in manufacturing. some other applications are computer-aided design (CAD), biometrics, and paleontology.

Graph matching enhances content-based ***image retrieval*** systems by enabling more accurate and semantically meaningful comparisons between images. Images are represented as scene graphs capturing objects and their relationships. Graph matching identifies similar graphs in a database. It can be used in retrieving images with similar spatial arrangements of objects and querying annotated images with semantic graph description. Other applications include digital libraries, e-commerce, and multimedia archives.

Graph matching is employed in ***motion analysis*** to track objects or regions across a sequence of frames. Nodes represent tracked points in consecutive frames, and edges encode spatial or temporal relationships. Matching identifies corresponding points across frames. It can be used in tracking human motion for activity recognition and monitoring vehicle movement in traffic analysis. Other applications of motion analysis are video surveillance, sports analysis, and robotics.

Graph matching helps in understanding the structural layout of an image or scene by identifying objects and their relationships. ***Scene graphs*** are generated to represent spatial or semantic relationships between objects. Matching enables comparison or alignment with known scenes. It helps in analyzing indoor scenes for autonomous robots. It is also used in detecting anomalies in architectural layout. Some other applications of scene analysis are smart homes, urban planning and robotics.

Biomedical image analysis use graph matching in analyzing and comparing complex biomedical structures represented as graphs. Nodes represent anatomical structures or features, and edges encode their spatial or functional relationships. Examples of biomedical image analysis are matching brain networks in neurological studies, Identifying cellular structures in microscopy images. Its applications are in neuroscience, pathology, and drug development.

Graph matching helps in ***fault detection*** by identifying defects in manufactured components by comparing their graph representation with defect-free models. In this scenario nodes represent features or regions of a component, and edges encode spatial relationships. Deviations in matching indicate defects. Some of the examples of fault detection are inspecting printed circuit boards (PCBs) for defects, ensuring consistency in assembly line products. Applications of fault detection are in quality control and automated inspection systems.

Graph matching offers a versatile and powerful framework for addressing various image analysis tasks. Its ability to model complex relationships and handle structural variability makes it indispensable in fields ranging from medical imaging to robotics. As advancements in algorithms and computational resources continue, graph matching is poised to play an even more significant role in solving challenging problems in image analysis.

Chapter 9

Graph Matching in Social Network Analysis

Social networks are complex structures that consist of individuals or entities (nodes) and the relationships or interactions (edges) between them. Analyzing these networks is critical for understanding social dynamics, community structures, information diffusion, and many other phenomena. Social Network Analysis (SNA) is the study of social structures through the use of network and graph theory. SNA is a powerful method for studying the structure and dynamics of relationships among entities such as individuals, organizations, or even entire societies. It uses graph theory to model these relationships and analyze patterns within networks, offering insights into social structures, communication flows, and the influence of actors within a network. It involves analyzing the patterns of connections between nodes (such as individuals, organizations, or countries) to understand the underlying social structure and dynamics. Graph matching is a powerful tool in SNA that enables the comparison, alignment, and integration of different networks. This chapter delves into the role of graph matching in social network analysis, exploring its techniques, applications, and challenges.

This chapter is organized as follows. Section 9.1 describes the basic concepts of social network analysis. It introduces various centrality measures used to quantify the importance of nodes within a network. Section 9.2 presents different methods of community detection in social network. Link prediction is discussed in Section 9.3. It also explains the important methods of link prediction along with the metrics to evaluate the performance of link prediction models. Section 9.4 presents the graph-based recommendation system in social network. Finally, Section 9.5 describes the influence analysis using graphs.

9.1 Basic Concepts in Social Network Analysis

SNA has roots in sociology, anthropology, and psychology. Early sociograms by Jacob Moreno in the 1930s visualized social relationships and laid the groundwork for modern network theory. The formalization of network analysis techniques, especially in the 1970s and 1980s, led to the

DOI: 10.1201/9781003532248-9

development of a robust framework for analyzing social structures. SNA involves the systematic study of social structures through networks and graph theory. The focus of SNA is on the relationships between entities (nodes) and the connections (edges) between them.

In social network ***nodes*** or vertices represent entities such as individuals, organizations, or concepts. ***Edges*** or links represent relationships or interactions between nodes. Edges can be directed (one-way) or undirected (two-way) and weighted or unweighted. ***Paths*** are sequences of edges that connect two nodes. ***Connectivity*** measures the ease with which nodes can reach each other. Figure 9.1 shows an example of social network which represents individuals as nodes and their connections as edges.

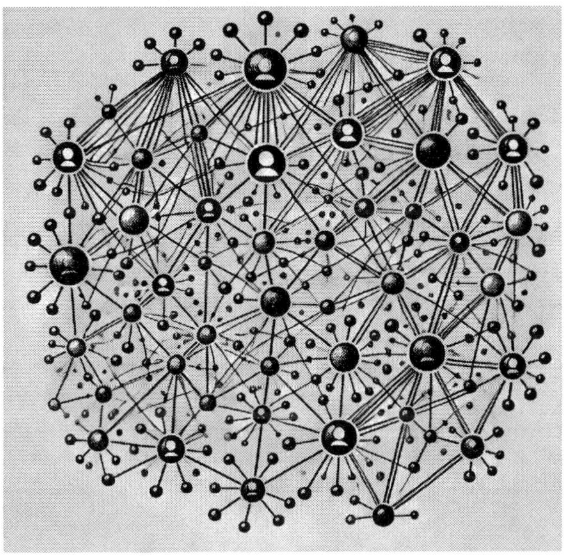

FIGURE 9.1: Social Networks

There are several types of social networks such as ego networks, global networks and multilayer networks. An ***Ego network*** is a type of social network that focuses on a single individual, known as the ego, and their direct connections, known as alters. It represents the local neighborhood of the ego within a larger network, including the links between alters. It focus on a single individual ego and the direct connection around them. Figure 9.2 shows an example of ego network, where black node represent ego and gray nodes surrounding it denotes alters. In ego networks the edges are the connections between the ego and alters, and potentially among the alters themselves. An example of an ego network is a person's friend circle on a social media platform, where the ego is the user, and alters are their friends.

Global network consider the entire network, analyzing all nodes and edges within the system. ***Multilayer network*** involve multiple types of

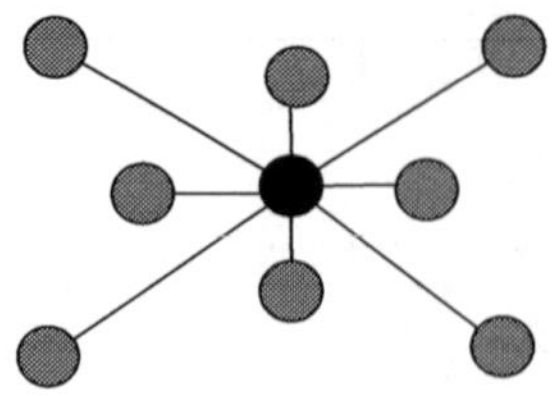

FIGURE 9.2: Ego Network

relationships or interactions, with different layers representing different type of connections.

9.1.1 Centrality Measures

Quantifying the structure of a network is central to SNA. Various metrics are used to describe and analyze the properties of networks. ***Centrality measures*** identify the most important or influential nodes in a network. A substantial body of research on networks has been dedicated to the study of centrality, a concept used to quantify the importance of nodes within a network. Since importance can be defined in various ways, numerous centrality measures have been introduced to capture different aspects of a node's influence [74].

One of the simplest and most widely used centrality measures is ***degree centrality***, which is defined as the number of edges connected to a given node.

$$C_D(v) = \frac{\deg(v)}{|V| - 1}$$

where $deg(v)$ is the number of edges connected to node v.

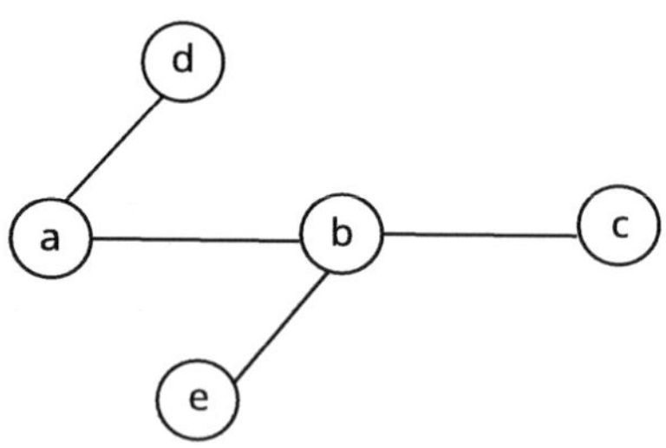

FIGURE 9.3: Graph

For the example graph shown in Figure 9.3, degree centrality of individual nodes are as follows. $C_D(a) = 2/4, C_D(b) = 3/4, C_D(c) = 1/4, C_D(d) = 1/4, C_D(e) = 1/4$. In the above graph, node b has highest degree centrality.

In the context of social networks, this measure is often explicitly referred to as degree centrality to highlight its relevance in assessing a node's prominence.

In directed networks, a node has both an in-degree (the number of incoming edges) and an out-degree (the number of outgoing edges), both of which can serve as meaningful indicators of centrality depending on the context.

Although degree centrality is a straightforward metric, it can provide valuable insights into network structure and dynamics. For instance, in social networks, individuals with a high number of connections are often assumed to have greater influence, wider access to information, or higher social prestige compared to those with fewer connections.

Beyond social networks, degree centrality is also widely applied in other domains. A notable example is its use in *citation networks*, where the in-degree of a scientific paper—defined by the number of times it has been cited by other papers—serves as a rough measure of its influence. Papers with a high number of citations are generally regarded as having made significant contributions to their field, making citation count a widely used metric for evaluating scientific impact.

Betweenness centrality measures how often a node lies on the shortest paths between other nodes.

$$C_B(v) = \sum_{s \neq v \neq t} \frac{\sigma(s,t|v)}{\sigma(s,t)}$$

where $\sigma(s,t)$ is the total number of shortest paths between s and t, and $\sigma(s,t|v)$ is the number passing through v.

Node b has highest betweenness centrality in Figure 9.3 as it connects multiple paths. Nodes with high-betweenness centrality play a crucial role in the network, as they act as intermediaries for information flow. Therefore, nodes with high-betweenness centrality can control information flow [73].

Consider a network where some entity—such as messages, data, or information—flows from one vertex to another along the edges. In a social network, for example, people may exchange messages, share news, or spread rumors. On the Internet, data packets travel between routers, following specific paths. To analyze the importance of different nodes in such a network, let us assume a simple model: each pair of connected vertices exchanges messages with equal probability per unit time, and messages always follow the shortest (geodesic) path between sender and receiver. If multiple shortest paths exist, one is chosen at random. Given this setup, an important question arises: on average, how many messages will pass through each vertex on their way to their destination?

The answer is straightforward: since messages travel along each geodesic path at the same rate, the number of messages passing through a vertex is proportional to the number of shortest paths that include that vertex. This measure is known as betweenness centrality, or simply betweenness.

The more geodesic paths a vertex belongs to, the more control it has over communication between other nodes. In practical scenarios, nodes with high-betweenness centrality may exert influence by facilitating or obstructing the spread of information. For instance, in a message-passing network, a highly central vertex could gain power by accessing messages that pass through it or by charging a fee for relaying them.

Additionally, removing a high-betweenness vertex can cause significant disruption in the network's communication, as it serves as a key bridge between different parts of the system. However, in real-world networks, communication patterns are often more complex. Not all nodes exchange information with equal frequency, and messages do not always follow the shortest path. Despite these complexities, betweenness centrality remains a useful approximation of a vertex's influence over information flow, making it a valuable tool for analyzing the structure and function of various types of networks.

Closeness centrality measures the average length of the shortest path from a node to all other nodes.

$$C_C(v) = \frac{1}{\sum_u d(v,u)}$$

where $d(v,u)$ is the shortest distance from v to u.

In Figure 9.3, node b has the highest closeness centrality since it has the shortest average distance to all other nodes.

Nodes with high-closeness centrality can quickly interact with others. A *shortest path*, also known as a ***geodesic path***, is the path between two vertices that has the minimum possible length. The length of this path, referred to as the ***geodesic distance*** or ***shortest distance***, represents the smallest number of edges required to connect the two vertices within the network. More formally, the geodesic distance between two vertices, i and j, is defined as the smallest integer r for which the entry $[A^r]_{ij} > 0$, where A is the adjacency matrix of the network. This condition ensures that a path of length r exists between i and j.

Let d_{ij} represent the length of a geodesic path from vertex i to j, which corresponds to the number of edges along the shortest path between them. The mean geodesic distance for vertex i denoted as ℓ_i, is the average geodesic distance from i to all other vertices in the network. Vertices with low mean geodesic distance are, on average, closer to other vertices, meaning they can access information more quickly or exert influence more directly over others. In a social network, for instance, an individual with a lower mean distance to others may find that their opinions, news, or influence spread more rapidly throughout the community compared to someone with a higher mean distance.

Unlike traditional centrality measures, the mean geodesic distance ℓ_i does not assign higher values to more central vertices—instead, central vertices have lower values, and less central vertices have higher values. To maintain consistency with other centrality measures, researchers in social network analysis typically use the inverse of ℓ_i instead of ℓ_i itself. This inverted measure is known as closeness centrality, denoted as:

$$C_i = \frac{1}{\ell_i} = \frac{n}{\Sigma_j d_{ij}}.$$

Closeness centrality quantifies how quickly a node can interact with others in the network. A higher closeness centrality value indicates that a vertex is, on average, more centrally located, allowing for more efficient communication and influence within the network.

Eigenvector centrality is a generalization of degree centrality which gives each node a value based on the sum of the values of its neighbors.

$$C_E(v) = \frac{1}{\lambda} \sum_{u \in N(v)} A_{uv} C_E(u)$$

where A_{uv} is the adjacency matrix, and λ is a scaling factor.

Eigenvector centrality refines the idea of measuring a vertex's importance based on its connections. While degree centrality assigns one centrality point for each direct neighbor, eigenvector centrality takes into account the importance of those neighbors as well. The fundamental idea is that a vertex is more influential if it is connected to other highly influential vertices rather than just many less important ones. Instead of simply counting the number of neighbors, eigenvector centrality assigns each vertex a score that is proportional to the sum of the centralities of its neighbors. This creates a recursive relationship where a node's importance is derived from the importance of its neighbors.

To compute eigenvector centrality, we start with an initial estimate for the centrality score of each vertex i, denoted as x_i. A common choice is to set $x_i = 1$ or all vertices initially. While this starting value is arbitrary, it provides a basis for an iterative refinement of centrality scores. The updated centrality score for vertex i, denoted as x'_i, is calculated as:

$$x'_i = \sum_j A_{ij} x_j$$

where A_{ij} represents the adjacency matrix of the network, meaning $A_{ij} = 1$ if there is an edge between vertices i and j, and 0 otherwise. This equation essentially states that a vertex's centrality is the sum of the centralities of its neighbors. By repeating this update process iteratively, the values will eventually converge to a stable set of scores, which correspond to the eigenvector associated with the largest eigenvalue of the adjacency matrix. Hence, the measure is called eigenvector centrality. For a node i its eigenvector centrality is given by

$$x_i = \kappa_1^{-1} \sum_j A_{ij} x_j$$

where κ_1 is the largest eigenvalue of adjacency matrix A_{ij}. Eigenvector centrality is particularly useful in identifying key influencers in networks, such as important web pages, influential people in social networks, or critical nodes in power grids or transportation systems.

9.1.2 Similarity Measures

In SNA, measuring the similarity between vertices (nodes) is crucial for tasks such as community detection, link prediction, recommendation systems, and network clustering. The similarity between two vertices can be computed based on different criteria, including structural, neighborhood, and path-based measures. Below, we categorize and describe the main techniques used to measure similarity in social networks.

Structural Similarity Measures

Structural similarity measures compare how similar two nodes are based on their direct connections or graph structure.

Common Neighbors (CN) measures the number of shared neighbors between two nodes.

$$S_{CN}(u, v) = |\Gamma(u) \cap \Gamma(v)|$$

Where $\Gamma(u)$ is the set of neighbors of node u. As an example of CN, in a co-authorship network, two researchers with many common co-authors have a higher similarity.

Jaccard Similarity measures the overlap between the neighborhoods of two nodes, normalized by the union of their neighborhoods.

$$S_{Jaccard}(u, v) = \frac{|\Gamma(u) \cap \Gamma(v)|}{|\Gamma(u) \cup \Gamma(v)|}$$

Jaccard similarity is used for link prediction in recommendation system.

Adamic-Adar Index (AA) weighs shared neighbors based on their degree (nodes with high degrees contribute less).

$$S_{AA}(u, v) = \sum_{w \in \Gamma(u) \cap \Gamma(v)} \frac{1}{\log d(w)}$$

where $d(w)$ is the degree of node w. It is applicable in social media friend recommendations.

Resource Allocation Index (RA) is similar to Adamic-Adar but assumes that smaller-degree nodes contribute more similarity.

$$S_{RA}(u, v) = \sum_{w \in \Gamma(u) \cap \Gamma(v)} \frac{1}{d(w)}$$

It can be used in predicting missing links in protein-protein interaction networks.

Path-Based Similarity Measures

These measures consider the shortest paths or multiple paths between two nodes in the network.

Shortest Path Similarity measures similarity based on the length of the shortest path between two nodes.

$$S_{SP}(u,v) = \frac{1}{\text{shortest path length}(u,v)}$$

If nodes are not connected, similarity is 0. It is used in social influence analysis where closer users have stronger influence.

Katz Index considers all paths between two nodes but gives shorter paths more weight.

$$S_{Katz}(u,v) = \sum_{l=1}^{\infty} \beta^l \cdot |\text{paths of length } l \text{ from } u \text{ to } v|$$

where β is a decay factor ($0 < \beta < 1$). Application of Katz index is link prediction in collaboration networks.

SimRank depicts two nodes are similar if they are linked to similar nodes.

$$S_{SimRank}(u,v) = C \cdot \frac{\sum_{i=1}^{|\Gamma(u)|} \sum_{j=1}^{|\Gamma(v)|} S(w_i, w_j)}{|\Gamma(u)| \cdot |\Gamma(v)|}$$

where C is a decay factor. It is used in web page similarity analysis.

Attribute-Based Similarity Measures

These techniques use node attributes (e.g., age, location, interests) to measure similarity.

Cosine Similarity measures similarity between feature vectors of two nodes.

$$S_{cosine}(u,v) = \frac{\mathbf{x}_u \cdot \mathbf{x}_v}{||\mathbf{x}_u||||\mathbf{x}_v||}$$

where $\mathbf{x}_u$ and $\mathbf{x}_v$ are attribute vectors of nodes u and v. Cosine similarity can be used in user-based collaborative filtering for recommendations.

Euclidean Distance measures distance between feature vectors.

$$S_{Euc}(u,v) = \frac{1}{1 + ||\mathbf{x}_u - \mathbf{x}_v||}$$

Euclidean distance can be applied for finding similarity in social media profiling.

Pearson Correlation Coefficient measures the linear relationship between numerical attributes.

$$\text{Sim}(A, B) = \frac{\sum(A_i - \bar{A})(B_i - \bar{B})}{\sqrt{\sum(A_i - \bar{A})^2} \cdot \sqrt{\sum(B_i - \bar{B})^2}}$$

Pearson correlation is used for rating-based systems such as movie recommendations.

Hamming Distance is used for attributes represented as binary strings, Hamming distance counts the number of differing attributes.

$$\text{Hamming}(A, B) = \sum_{i=1}^{n} |A_i - B_i|$$

Hamming distance is used for binary attributes such as presence or absence of traits. Attribute-based similarity in social networks helps in friend recommendations, community detection, link prediction, and recommender systems. Various similarity measures like Jaccard, Cosine, Euclidean distance, and Pearson correlation can be applied based on the type of attributes.

Hybrid Similarity Measures

It combining multiple techniques improves accuracy in real-world networks.

Hybrid graph and attribute-based similarity combines structural measures (e.g., Jaccard similarity) with attribute-based methods (e.g., Cosine similarity). As an example friend suggestion algorithms in Facebook and LinkedIn use hybrid and attribute-based similarity.

Machine Learning-Based Similarity uses supervised learning models (e.g., Random Forest, Neural Networks) trained on multiple similarity features. Traditional similarity measures (e.g., Jaccard, Cosine, Euclidean distance) rely on handcrafted formulas to quantify the similarity between users or entities in a social network. However, these approaches may not fully capture complex, non-linear relationships in large and dynamic social graphs. Machine learning-based similarity measures offer a more adaptive and data-driven approach by learning patterns from data rather than relying on predefined formulas. These methods leverage supervised, unsupervised, and deep learning techniques to enhance similarity estimation in tasks like friend recommendations, link prediction, and community detection. Examples of machine learning-based similarity include fraud detection and anomaly detection in social networks. Table 9.1 lists the applications of machine learning-based similarity.

TABLE 9.1: Applications of Machine Learning-Based Similarity

Application	Machine Learning Approach
Friend Recommendation	Supervised Learning
Community Detection	Unsupervised Learning
Link Prediction	Graph Embeddings
Recommender Systems	Matrix Factorization, GNNs
Influence Detection	GNNs, Siamese Networks

Machine learning techniques for computing similarity in social networks can be broadly divided into following categories.

1. *Supervised Learning-Based Similarity* trains a model on labeled data. It learns feature importance and relationships between users. Example includes classification models predicting whether two users are similar.

2. *Unsupervised Learning-Based Similarity* are cluster users based on their attributes and behavior. For example, clustering users with similar interests without predefined labels.

3. *Graph-Based and Deep Learning Similarity* uses graph embeddings and neural networks to compute similarity. Graph-based methods learn from network structure to model user similarity.

Despite challenges like sparsity and high dimensionality, advancements in machine learning and graph-based models continue to improve similarity estimation in social networks. Measuring similarity between vertices in social networks is fundamental for tasks like community detection, recommendation systems, and link prediction. The choice of similarity measure depends on the network structure, available attributes, and application domain.

9.2 Community Detection in Social Network

Community detection is a fundamental task in SNA, aiming to identify groups or clusters of nodes (individuals or entities) that are more densely connected internally than with the rest of the network. These communities often represent meaningful structures such as groups of friends, professional networks, or communities sharing common interests [35].

A community is a subset of nodes within a network that are densely connected with each other but sparsely connected with the rest of the network. The definition of "dense" and "sparse" connections varies based on the algorithm or approach used. Social networks are modeled as graphs, with nodes representing individuals and edges representing interactions, relationships, or communication between them. Communities can be overlapping as well as non-overlapping. In overlapping communities nodes can belong to multiple communities such as a person in multiple social group, while in non-overlapping communities nodes belong to only one community.

Community detection helps in understanding network structure by revealing the underlying organizational principles of a network. It also helps in predictive analysis by predicting user behavior, interests or information speed. For example, it detects influential individuals who can spread information efficiently within a community. It can also be used to identify malicious groups, fake accounts, and suspicious interactions. Some of the major applications of community detection in SNA are in marketing and recommendation systems, fraud detection, public health for tracking the spread of diseases and enhancing communication strategies.

9.2.1 Methods of Community Detection

Community detection is an essential task in analyzing social networks, as it helps uncover groups of nodes (users, entities) that are more densely connected to each other than to the rest of the network. Several methods have been developed to identify communities in graphs, each with its strengths and limitations. Below, we discuss the major categories of community detection techniques.

Graph Partitioning Methods

Graph partitioning techniques divide a network into non-overlapping subsets by optimizing certain criteria, such as minimizing the number of edges between groups. *Minimum Cut Partitioning* splits the graph by minimizing the number of edges that need to be removed to separate the network into predefined groups. But it requires the number of communities to be predefined, which is often unknown. *Spectral clustering* uses the eigenvalues of the graph Laplacian matrix to perform clustering. It converts the graph into a low-dimensional space where standard clustering techniques (e.g., K-means) can be applied. This method works well for balanced communities. It is computationally expensive for large graphs.

Modularity-Based Methods

Modularity is a measure that quantifies the quality of a network partition by comparing the density of connections inside communities to that of a random graph. *Louvain* method is based on modularity-based approach [17]. This method starts with each node as a separate community. It iteratively merges communities to maximize modularity. This method use a hierarchical approach, where small communities are merged into larger ones. Louvain method is very efficient for large-scale networks but it may not detect small communities accurately.

Girvan-Newman algorithm is another algorithm based on modularity-based method [81]. It measures edge betweenness centrality which is the number of shortest paths passing through an edge. Then it iteratively removes edges with the highest betweenness, breaking the graph into communities. This algorithm works well for small, well-separated communities. It may be computationally expensive for large networks.

Clustering-Based Methods

Clustering-based methods treat nodes as points in space and apply traditional clustering techniques. One such method is k-means clustering which converts graph nodes into vector representations (e.g., adjacency-based features). It applies k-means clustering to group similar nodes. While this method is simple and scalable it requires the number of communities to be predefined. Density-based clustering is another approach which group nodes based on density instead of distance, allowing for arbitrary-shaped communities. This method does not need the number of communities to be predefined. One of its limitations is that it is difficult to determine the correct density threshold.

Overlapping Community Detection

Traditional methods assume that each node belongs to only one community, but in real-world social networks, nodes (e.g., people) often belong to multiple groups. *Clique percolation method* identifies fully connected subgraphs or cliques. It merges overlapping cliques to form communities. This method captures overlapping communities. *Fuzzy clustering* is another method which assigns each node a probability of belonging to multiple communities. This approach is more realistic for social networks. On the other hand it is harder to interpret the result in this method.

Probabilistic and Statistical Approaches

These methods model community structures using probabilistic techniques. Methods in this category utilize probabilistic approaches to infer the distribution of community memberships for nodes in a node-attributed social network. This inference is based on the assumption that both the network structure and node attributes are generated according to predefined parametric distributions. Typically, generative and discriminative models are employed for this purpose. However, selecting appropriate prior distributions for the network structure and attributes is a challenging and non-trivial task, as it significantly impacts the accuracy and effectiveness of the community detection process. ***Stochastic Block Model*** (SBM) assumes that nodes belong to hidden communities and edges are generated based on probability distributions.

TABLE 9.2: Comparison of Community Detection Methods

Method	Overlapping	Scalability
Graph Partitioning	No	Moderate
Modularity-Based	No	High
Edge Betweenness	No	Low
K-Means Clustering	No	High
Clique Percolation	Yes	Low
Stochastic Block Model	Yes	Low
Graph Neural Networks	Yes	Moderate

SBM provides a theoretical foundation for community detection. This approach is computationally demanding for large networks. ***Mixed Membership Models*** are similar to SBM but allows nodes to belong to multiple communities with different probabilities [3]. Table 9.2 lists the comparison of various community detection methods.

Deep Learning-Based Approaches

Deep learning has emerged as a powerful tool for community detection in social networks, offering more flexibility and accuracy than traditional graph-based methods. These approaches leverage neural networks to learn latent representations of nodes and edges, enabling the discovery of complex, hierarchical, and overlapping community structures. Recent advancements in Graph Neural Networks and embedding-based techniques have improved community detection.

Graph Neural Networks (GNNs) are a class of deep learning models designed to process graph-structured data. They use neural layers to aggregate information from a node's neighbors, learning representations that capture community structures. ***Graph Convolutional Networks*** (GCNs) extend convolutional neural networks (CNNs) to graph data. They aggregate information from neighboring nodes using spectral graph theory. In these approaches nodes are embedded into a low-dimensional space. Clustering techniques are applied to group nodes into communities. They captures both node attributes and network structure. GNNs and GCNs are particularly suitable for large-scale networks. ***Graph Attention Networks*** (GATs) improve upon GCNs by using attention mechanisms to assign different weights to different neighbors, allowing the model to focus on important node connections. It learns node embeddings while preserving local and global community structures. GATs are more flexible than GCNs in handling heterogeneous networks. It is also suitable for dynamic and evolving communities. ***Graph Autoencoders*** (GAEs) learn a compressed representation of a graph by encoding node features and reconstructing the graph structure. ***Variational Graph***

TABLE 9.3: Comparison of Deep Learning-Based Community Detection Methods

Method	Overlapping	Scalability	Best for
GCNs	No	High	Large-scale networks
GATs	No	Moderate	heterogeneous networks
GAEs	Yes	Moderate	Unsup. community detection
VGAEs	Yes	Moderate	Unsup. community detection
DeepWalk	No	High	Large graphs
Node2Vec	Yes	High	Structural similarity
SDNE	Yes	Moderate	Complex graphs
Graph RL	Yes	Low	Dynamic communities

Autoencoders (VGAEs) extend GAEs by using probabilistic inference, allowing for uncertainty estimation. In this technique node are embedded in a latent space. Communities are formed based on clustering in the learned representation. This approach is effective for detecting overlapping and hierarchical communities. It works well for unsupervised learning.

Node embedding techniques map graph nodes into a continuous vector space while preserving community structures. *DeepWalk* uses random walks to generate node sequences, treating the network like a text corpus. It applies a popular NLP technique *word2vec* to learn embedding. Nodes with similar embeddings are clustered together. This technique is simple and efficient for large networks. It captures both local and global community structures. *Node2Vec* is an extension of DeepWalk that introduces biased random walks to balance between local (homophily) and global (structural similarity) node relationships. It is similar to DeepWalk, here embedding are clustered to detect communities. This technique is more flexible than DeepWalk. It can identify both overlapping and non-overlapping communities. ***Structural Deep Network Embedding*** (SDNE) uses deep autoencoders to learn node representations while preserving graph structure. It perform clustering on learned node embeddings. SDNE preserves both local and global graph structure. It works well for complex and large-scale networks.

Reinforcement learning (RL) methods train agents to explore and partition graphs based on rewards for well-defined communities. ***Graph Reinforcement Learning*** (GRL) uses an RL agent to explore the network and assign nodes to communities based on learned policies. GRL approaches adapts well to dynamic and evolving communities. It can handle overlapping communities effectively.

Community detection in social networks is a crucial task for understanding user interactions and network structures. Different techniques are suitable for different network types, ranging from modularity-based approaches like Louvain (for large-scale networks) to deep learning-based methods (for complex and overlapping communities). Table 9.3 lists the comparison of various deep-learning-based community detection methods. It compares methods with whether it can handles overlapping communities, its scalability and one of its best usage. The choice of method depends on factors such as scalability, overlapping structures, and computational efficiency.

9.3 Link Prediction

Link prediction and recommendation systems are two important tasks in the analysis of social networks. They aim to understand and predict relationships or suggest connections, respectively, by leveraging the underlying structure and attributes of the network.

Link prediction seeks to predict the likelihood of future or missing connections (edges) between nodes in a social network based on the current network structure and available attributes. It can be used to predict potential friendships in social media platforms. Link prediction helps to identify collaborative networks of potential collaborators in research or professional networks. It can also be used in fraud detection to identify suspicious or unusual connections in financial or communication networks.

9.3.1 Link Prediction Methods

9.3.1.1 Structural Methods

These methods rely on predefined similarity measures to determine the likelihood of a link forming between two nodes.

Common Neighbors (CN)
The more mutual friends two users share, the higher the probability of forming a link.

$$CN(x, y) = |\Gamma(x) \cap \Gamma(y)|$$

where $\Gamma(x)$ represents the set of neighbors of node x. Used for social media friend recommendations.

Jaccard Coefficient
Measures the proportion of common neighbors relative to the total number of neighbors.

$$J(x, y) = \frac{|\Gamma(x) \cap \Gamma(y)|}{|\Gamma(x) \cup \Gamma(y)|}$$

used for predicting collaborations in academic networks.

Adamic-Adar Index
Gives more weight to shared neighbors that have fewer overall connections.

$$AA(x, y) = \sum_{z \in \Gamma(x) \cap \Gamma(y)} \frac{1}{\log |\Gamma(z)|}$$

used for e-mail and communication networks.

Preferential Attachment (PA)
The probability of a new connection is proportional to the product of node degrees.

$$PA(x, y) = |\Gamma(x)|.|\Gamma(y)|$$

Used for growth models for online social networks.

Resource Allocation Index (RA)
Similar to Adamic-Adar but with a stronger penalty for high-degree nodes.

$$RA(x, y) = \sum_{z \in \Gamma(x) \cap \Gamma(y)} \frac{1}{|\Gamma(z)|}$$

Used for biological network link prediction.

9.3.1.2 Probabilistic and Graphical Models

Probabilistic models for link prediction leverage probabilistic principles to model uncertainty in relationships and infer missing links based on observed data. These models estimate the probability of a link existing between two nodes by utilizing structural, node-level, and edge-level features. These methods use probability distributions to estimate the likelihood of link formation.

Bayesian Probabilistic Models

Bayesian models estimate the probability of a link based on prior knowledge and observed data, updating beliefs as new information becomes available. *Bayesian Network Models* represent dependencies between nodes and links using a directed acyclic graph (DAG). It incorporate prior probabilities and use Bayesian inference to update link probabilities. For example, if two authors have co-authored papers before, a Bayesian model may predict future collaborations based on prior co-authorship patterns. *Naïve Bayes Classifiers* assume independence between features and use Bayes' theorem to compute the probability of a link existing. For example, If a node has a high degree and many mutual neighbors with another node, the model assigns a higher probability to a potential link.

Markov Models and Probabilistic Graphical Models

Markovian models assume that the probability of a link depends on the immediate neighborhood of nodes. ***Markov Random Fields*** (MRF) uses an undirected graphical model where the probability of a link depends on its local neighborhood. It can be used in social networks where friendships are influenced by mutual friends and local connectivity. ***Hidden Markov Models*** (HMMs) is useful for predicting temporal link formation in dynamic networks. It can predict links in evolving citation networks by modeling how research topics evolve over time.

9.3.1.3 Probabilistic Graph Embedding Models

Graph embedding models represent nodes as probabilistic latent vectors in a continuous space, estimating link probabilities based on distance or similarity in the embedding space. ***Deep Probabilistic Graph Embeddings*** use deep learning models, such as variational autoencoders (VAE) or generative adversarial networks (GANs), to learn probabilistic representations of nodes. Variational graph autoencoders learn a latent representation of nodes and predict missing links as probabilistic reconstruction. ***Stochastic Block Models*** (SBM) assume nodes belong to latent communities and links are formed based on community structure. It can predict collaborations in academic networks where authors belong to different research communities.

There are several advantages of probabilistic models for link prediction. It handles uncertainty by providing probabilistic confidence scores for predicted links, making them useful in uncertain environments. Bayesian models and Markov models incorporates prior knowledge by integrating domain knowledge for better predictions. Probabilistic models such as graph embedding model are scalable to large networks. These models can easily adapt to different type of networks including social, biological and citation networks.

Probabilistic models for link prediction provide a robust framework for handling uncertainty and making informed predictions in complex networks. They leverage various mathematical and machine learning principles, including Bayesian inference, Markovian assumptions, and latent feature modeling, to estimate link probabilities. Their flexibility and scalability make them essential tools in social network analysis, modern network science and data-driven applications.

9.3.2 Evaluation Metrics

Evaluating the performance of link prediction models requires appropriate metrics to measure how well the predicted links align with the actual connections in the network. Below are the most commonly used evaluation metrics.

Ranking-Based Metrics: These metrics evaluate how well the model ranks the actual links among all possible links.

Area Under the Receiver Operating Characteristic Curve (AUC)
Measures the probability that a randomly chosen positive link (existing link) is ranked higher than a randomly chosen negative link (non-existing link).

$$AUC = \frac{\text{Number of correct positive-negative link orderings}}{\text{Total number of positive-negative pairs}}$$

Range: $0 \leq AUC \leq 1$, where 0.5 represents random prediction and 1 indicates perfect ranking.

Mean Reciprocal Rank (MRR)
Evaluates how high the first correct predicted link appears in the ranked list. It is useful when prioritizing the most relevant link predictions.

$$MRR = \frac{1}{N}\sum_{i=1}^{N}\frac{1}{\text{rank}_i}$$

Classification-Based Metrics: These metrics treat link prediction as a binary classification problem, where predicted links are compared to actual links.

Precision
Measures the proportion of correctly predicted links among all predicted links.

$$\text{Precision} = \frac{\text{True Positives (TP)}}{\text{TP + False Positives (FP)}}$$

Recall
Measures the proportion of correctly predicted links among all actual links

$$\text{Recall} = \frac{\text{TP)}}{\text{TP + False Negatives (FN)}}$$

F1-score
Harmonic mean of Precision and Recall, balancing both metrics. Best for evaluating when predicting missing or future links is critical.

$$F1 = 2 \times \frac{\text{Precision} \times \text{Recall}}{\text{Precision} + \text{Recall}}$$

Accuracy
Measures the proportion of correctly predicted links among all possible links.

$$\text{Accuracy} = \frac{\text{TP + True Negatives (TN)}}{\text{Total Samples}}$$

Matthews Correlation Coefficient (MCC)
A balanced measure even for imbalanced datasets.

$$MCC = \frac{TP \times TN - FP \times FN}{\sqrt{(TP + FP)(TP + FN)(TN + FP)(TN + FN)}}$$

It ranges from -1 (worst) to +1 (best), where 0 means no better than random guess.

Network-Specific Metrics: These metrics evaluate link predictions in the context of network properties. Structural consistency measures how well the predicted links align with the existing network structure. Network connectivity assesses whether predicted links help maintain or enhance the connectivity of the network, ensuring realistic link formations.

Choosing the right evaluation metric depends on the application. Ranking-based metrics are ideal for recommendation systems. Classification metrics are useful for applications where correctly identifying missing links is crucial. Network-specific metrics ensure that predictions maintain the structural integrity of social networks.

9.4 Recommendation System

Recommendation systems in social networks help users discover relevant content, connections, and interactions based on their preferences, behaviors, and relationships with others. These systems leverage various data sources, including user activity, friendships, and shared interests, to provide personalized recommendations. Social networks generate massive amounts of data, making it challenging for users to find relevant information. Recommendation systems address this challenge by analyzing patterns in user behavior and suggesting items such as friends, groups, pages, advertisements, or content. Recommendation system is often used in social media for suggesting friends, groups, or pages to follow for different users. In e-commerce it helps recommending products or services to users.

9.4.1 Types of Recommendation Systems

Recommendation systems in social networks include graph-based recommendation, collaborative filtering, content-based filtering and hybrid recommendation system.

Graph-based recommendation systems are effective for recommendation as social networks are inherently structured as graphs (nodes represent users and edges represent connections). Friend recommendation uses graph traversal techniques to suggest connections. Community-based recommendation identifies groups or clusters within the network and recommends items relevant to the user's community. For example, LinkedIn suggests professional events attended by people in the same industry.

Collaborative Filtering (CF) methods recommend items based on the preferences and behaviors of similar users. It is widely used in friend recommendations, content suggestions, and product recommendations in social media. User-based collaborative filtering identifies users with similar preferences and recommends items that one user has liked but the other has not yet seen. Item-based collaborative filtering finds similar items based on user interactions and recommends them to users who liked similar items. As an example LinkedIn suggests professional groups based on the ones a user has already joined. The limitation of this approach is that it struggles to make recommendations for new users with little historical data.

Content-Based Filtering (CBF) recommends items by analyzing their features and matching them with user preferences. It relies on the similarity between items rather than user behavior. Feature-based matching uses keywords, categories, or descriptions to find similar content. For example, Instagram recommends posts based on hashtags and captions that a user frequently engages with. Profile learning builds a user profile based on past interactions and recommends items that match the user's interests. As an example, Twitter suggests accounts to follow based on the topics a user frequently interacts with.

Hybrid Recommendation Systems uses hybrid models to combine collaborative filtering and content-based filtering to improve recommendation accuracy and overcome individual limitations. Hybrid models can combine CF and CBF predictions to make final recommendation. It can use content-based methods for new user recommendation problem in collaborative filtering. For example, YouTube can recommends videos based on both viewing history (content-based) and trending videos among similar users (collaborative filtering).

Deep Learning-Based Recommendation models leverage neural networks to analyze complex user interactions and improve recommendation accuracy. Neural collaborative filtering uses deep neural networks to learn latent factors in user-item interactions. Such as TikTok's "For You" page continuously adapts recommendations based on user engagement. GNNs enhance graph-based recommendations by capturing multi-hop relationships between users and content. Twitter's recommendation engine uses GNNs to analyze the influence of social connections.

9.5 Influence Analysis using Graphs

Influence analysis in social networks involves studying how information, behaviors, or innovations spread within a network and identifying key individuals or nodes that significantly affect this propagation. Graph theory provides a powerful framework for modeling and analyzing such phenomena.

Influence measures the ability of a node to spread information or affect other nodes in the network. Influence can be local (affecting immediate neighbors) or global (spreading widely across the network). Influence can be direct such as a celebrity endorsing a product or indirect such as a message spreading virally across multiple connections. Influence analysis is widely used in marketing, politics, epidemiology, and misinformation detection.

Key aspects of influence analysis include identifying ***influential nodes*** for finding the most influential users in a network based on metrics like centrality, connectivity, and engagement. After identification, next stage is ***modeling influence spread*** to understand how information propagates through social interactions using models such as the Independent Cascade Model (ICM) and Linear Threshold Model (LTM). After modeling next step is to ***measure influence strength*** by evaluating how strongly a node affects others using graph-based measures like degree centrality, betweenness centrality, PageRank, and eigenvector centrality. Final aspect of influence analysis is to ***predict influence outcome*** by forecasting how changes in the network affect the spread of influence.

9.5.1 Models of Influence Spread

Influence spread models in social networks help understand how information, opinions, behaviors, or trends propagate through connections between individuals (nodes). These models are widely used in viral marketing, epidemiology, political campaigns, and misinformation detection. There are two primary categories of influence spread models. In ***deterministic models*** influence spread follows a fixed rule. Whereas in ***probabilistic models*** influence spread occurs with a certain probability such as cascade models.

Independent Cascade Model (ICM) is a probabilistic model of influence spread in which each active node has a chance to influence its inactive neighbors in discrete steps. At the start, an initial set of seed nodes is activated. Each activated node attempts to activate its neighbors with a probability p. If successful, the newly activated nodes try to influence their neighbors in the next step. The process continues until no more nodes can be activated. In this method influence spreads like a chain reaction.

Linear Threshold Model (LTM) is a deterministic, threshold-based model of influence spread in which a node is influenced if a sufficient fraction of its neighbors are already influenced. In this method each node u is assigned a randomly chosen threshold θ_u. At each time step, if the sum of influences from its active neighbors exceeds the threshold, the node becomes active. The process continues iteratively until no more activations occur. In this model influence spreads in a contagion-like manner, where peer pressure plays a role. An example of this model is a person adopting a new social media platform only if enough friends are already using it.

Susceptible-Infected-Recovered (SIR) Model is a probabilistic model used in epidemiology and information diffusion. This method models the spread of diseases, rumors, or information with three possible states for each node:

- Susceptible (S) – Not yet influenced.

- Infected (I) – Currently influenced and spreading information.
- Recovered (R) – No longer spreading (immune or uninterested).

A susceptible node becomes infected with probability p when in contact with an infected node. An infected node recovers with probability q after a certain time. The infection spreads until no more susceptible nodes remain.

Susceptible-Infected-Susceptible (SIS) Model is a probabilistic, cyclic model of influence spread. This model is similar to SIR, but recovered nodes can become susceptible again. The influence does not permanently disappear, and individuals may re-engage with the trend.

Heat Diffusion Model is a deterministic model of influence spread inspired from Physics. In this model influence spreads like heat conduction, where nodes with high influence diffuse it gradually to neighbors. Nodes with higher influence act as "hotspots" spreading influence across the network over time. Influence levels decrease as they move outward, similar to heat dissipating.

Game-Theoretic Influence Models are strategic and decision-making based model. It models influence spread as a game, where individuals make strategic choices based on incentives. For example, companies compete to maximize product adoption among consumers by offering discounts or rewards.

Understanding influence spread models helps in designing effective strategies for marketing, controlling misinformation, and predicting trends. Measuring influence strength in a social network involves evaluating how much impact a node (user) has on others. Various graph-based centrality measures, such as degree centrality, betweenness centrality, closeness centrality, eigenvector centrality etc., can used for this purpose.

Influence analysis provides valuable insights into how social networks function and helps optimize strategies for communication, marketing, and information control. Influence analysis is a critical area of study in social networks, offering insights into the dynamics of information propagation and providing actionable strategies for various applications, from marketing to public health. By leveraging graph-theoretical models and computational techniques, researchers and practitioners can better understand and harness the power of influence in complex networks.

Chapter 10

Recent Advances and Future Directions

This book addresses the critical problem of structural pattern recognition and graph matching, serving as a comprehensive resource for researchers, practitioners, and students interested in these fields. Graph matching is a fundamental process in pattern recognition that involves measuring the similarity between two graphs. Depending on the application and requirements, graph matching can be classified into two primary categories: exact graph matching and inexact graph matching. Exact graph matching is used when a precise one-to-one correspondence between two graphs is required. In contrast, inexact graph matching allows for flexibility and tolerance to errors, making it suitable for real-world applications where data may contain noise or distortions.

Graph matching plays a crucial role in various domains, including structural pattern recognition, biometric identification, computer vision, bioinformatics, cheminformatics, and social network analysis. The key advantage of using graphs instead of vectors for object recognition lies in their representational power and their flexibility in encoding relationships between multiple objects. However, this flexibility also introduces challenges. One of the most significant issues in graph matching is its high-computational complexity, making it computationally expensive for large-scale problems. Due to the lack of efficient exact solutions, several approximation and suboptimal algorithms have been developed. Another major challenge is the absence of standardized similarity measures to quantify the similarity or dissimilarity between two graphs accurately.

An extensive survey of exact and inexact graph matching techniques is presented in this book. One of the most widely used approaches for graph matching is graph edit distance (GED), which provides a flexible mechanism for comparing graphs. GED is defined as the minimum number of edit operations required to transform one graph into another, where the allowable operations include insertion, deletion, and substitution of nodes and edges.

A novel variation of graph edit distance is described using the concept of graph homeomorphism. The homeomorphic graph edit distance between two graphs is defined as the graph edit distance after applying path contraction operations to both input graphs. Path contraction simplifies the structure of graphs by removing nodes of degree two along simple paths, reducing

DOI: 10.1201/9781003532248-10

the search space and significantly improving computational efficiency. This approach helps address the computational challenges of graph matching by simplifying the graph representation while preserving the essential structural information.

A new class of graph matching algorithms is described, which reduces the graph size by eliminating less important nodes based on relevance measures. One such approach is the node contraction technique, where nodes with the lowest degree are removed, provided they are not cut vertices (i.e., their removal does not fragment the graph into disconnected components). This technique ensures that the structural integrity of the graph is maintained while achieving a balance between execution time and classification accuracy. The extended graph edit distance is then defined as the graph edit distance after applying appropriate node contraction operations to both graphs. Experimental results demonstrate that this approach effectively reduces computation time while maintaining classification performance.

A limitation of the node contraction approach is that the number of nodes of a given degree can vary across different graphs, leading to inconsistent reductions. To address this issue, an enhanced method is outlined, where the graph size is reduced by removing a fraction of nodes from both graphs based on a selected centrality measure. The four centrality measures explored in this framework include: degree centrality (importance based on node connections), betweenness centrality (importance based on shortest paths passing through a node), eigenvector centrality (importance based on connections to highly connected nodes), PageRank centrality (importance based on iterative ranking algorithms).

Experimental results reveal that different centrality measures lead to varied trade-offs between computational efficiency and classification performance. This method enables users to tailor the selection of the centrality measure based on specific application requirements. Furthermore, it allows for early estimation of graph matching results by focusing only on the most significant nodes, making it particularly beneficial for large-scale applications where real-time processing is required.

A new graph similarity metric is elucidated for measuring the similarity between two geometric graphs. This metric is designed to be both efficient and intuitive, capturing both structural and spatial relationships between graphs. Vertex distance is defined using a linear sum assignment problem, where nodes from one graph are optimally assigned to nodes of another to minimize overall distance. Edge distance is based on three fundamental edge features—orientation, length, and position. Graph Distance is defined as a composite measure integrating vertex distance with the three edge distance components to quantify overall similarity. This similarity measure provides a standardized approach to comparing geometric graphs and is particularly useful in applications such as computer vision, medical imaging, and geographic information systems.

The similarity framework is utilized for both exact and error-tolerant graph matching. An algorithm for geometric graph isomorphism is outlined to determine whether two graphs are structurally identical while preserving their geometric properties. Furthermore, a t-tolerant isomorphism algorithm is introduced, which allows for small distortions in vertex coordinates, making it robust to minor variations in spatial positioning. To enhance the reliability of geometric graph isomorphism, a graph alignment algorithm is employed before checking for isomorphism, ensuring that reference coordinates are optimally aligned.

Above error-tolerant graph matching framework extends the application of the geometric similarity metric to real-world scenarios involving large graphs with distortions. By using a set of weight parameters, users can adjust the relative importance of vertex distance, angular similarity, edge length, and positional similarity based on application needs. This flexibility enables the framework to handle diverse datasets, including biometric recognition, molecular structure analysis, and shape matching in computer vision.

The methodologies presented in this book offer innovative solutions to the fundamental challenges in structural pattern recognition and graph matching. The introduction of homeomorphic graph edit distance, node contraction techniques, centrality-based reductions, and geometric graph similarity measures provides a robust foundation for efficient and scalable graph matching algorithms. The proposed approaches are applicable to a wide range of domains, demonstrating the potential for real-world impact in fields such as computer vision, bioinformatics, and network analysis. This work contributes towards the development of standard similarity measures and computationally efficient solutions for large-scale graph matching problems, paving the way for future advancements in the field.

10.1 Emerging Trends

Emerging trends in structural pattern recognition and graph matching reflect advancements in machine learning, deep learning, optimization techniques, and computational efficiency. Below are some key trends shaping the future of this field.

Deep learning for graph matching and structural pattern recognition particularly Graph Neural Networks (GNNs) is one of the promising area. The rise of GNNs has revolutionized graph-based pattern recognition by enabling learning from structured data. Techniques such as Graph Convolutional Networks (GCNs) and Graph Attention Networks (GATs) improve the ability to model relational data effectively. Traditional graph matching methods rely on handcrafted similarity measures, but deep learning models now enable learning feature representations directly from raw data, reducing the need for manual tuning. Self-supervised learning approach help train models without extensive labeled data, making structural pattern recognition more scalable.

Many real-world graphs evolve over time (e.g., social networks, molecular structures). Dynamic and adaptive graph matching techniques that incorporate temporal changes are gaining attention. With increasing data complexity, efficient algorithms for large-scale graph matching are being developed. Techniques such as hierarchical matching, approximate nearest neighbor searches, and quantum computing applications help tackle computational challenges.

Advanced graph kernels incorporating deep learning techniques and higher-order structures provide more robust similarity measures. These methods improve tasks like classification, clustering, and retrieval. Graph embedding techniques transform graphs into continuous vector spaces which facilitates efficient computation and integration with deep learning models. Graph embedding methods, such as DeepWalk, Graph Autoencoders, enable better structural representation.

As AI-driven techniques become more prevalent, the need for interpretability increases. Researchers are focusing on explainable graph matching models that provide insights into decision-making processes. Interactive systems that allow domain experts to refine graph matching results ensure higher accuracy in critical applications, such as medical diagnostics and forensic analysis.

Combining graph-based approaches with image, text, and sensor data is an emerging trend in cross-domain areas like medical imaging, biometric recognition, and autonomous systems. Graph matching techniques are being applied across different domains, such as protein-protein interaction networks, social media analysis, and cybersecurity. Quantum computing offers potential speedups for combinatorial optimization problems in graph matching. Research into quantum annealing and quantum-inspired algorithms is growing. Brain-inspired computing architectures may offer energy-efficient solutions for real-time graph-based pattern recognition.

These emerging trends indicate that structural pattern recognition and graph matching are evolving towards more intelligent, scalable, and interpretable solutions, enabling their application in increasingly complex and dynamic domains.

10.2 Future Directions

Future research in structural pattern recognition and graph matching can explore several directions to enhance the current methodologies. One potential avenue is the evaluation of additional centrality measures across various datasets. The existing experimental results have been primarily derived from letter and molecular graph datasets. Future studies could extend these evaluations to other graph datasets, particularly in the domains of chemical and biological structures, where graph representations play a crucial role.

Another promising direction involves the implementation of parallelized versions of the presented algorithms. By leveraging parallel computing techniques, researchers can assess improvements in efficiency and scalability compared to traditional, sequential implementations. This can be especially valuable when dealing with large-scale graph datasets, where computational complexity poses a significant challenge. In the error-tolerant geometric graph matching framework, the current method for equalizing graph sizes before vertex distance computation involves inserting nodes into the smaller graph at coordinates equal to the average of all existing coordinates. Future research could investigate alternative strategies for node insertion, ensuring that the added nodes better represent the structural and geometric properties of the original graphs. Similarly, different methodologies could be explored for edge augmentation to maintain consistency between graph structures while preserving their intrinsic characteristics.

The current framework utilizes weight parameters to integrate vertex distance with the three components of edge distance: orientation, length, and position. Further experimental analyses could be conducted to explore different weight parameter combinations, assessing their impact on performance and optimizing the balance between accuracy and computational efficiency. Additionally, a significant direction for future work involves applying the geometric graph similarity framework to the development of graph kernels and embedding techniques. By leveraging the structured similarity measures provided by the framework, novel graph-based machine learning models can be designed for improved classification, clustering, and retrieval tasks in various domains. The integration of the framework with deep learning models and neural networks could further expand its applicability, paving the way for more robust and scalable solutions in structural pattern recognition and graph matching.

Chapter 11

Appendix: Graph Matching Tools

This appendix describes some of the important software toolkit and links for graph matching. Technical committee #15[1] of the International Association for Pattern Recognition[2] (IAPR) is devoted to encouraging the research and innovation in graph-based representation in the pattern recognition field. It provides details of various software toolkits, links and documentation useful for working in graph matching and graph-based representations. It also provides links for various benchmarking and graph datasets[3].

Various data structures and graph types are proposed to represent different types of graphs. Holt *et al.* [95] in 2000 describe one of the important structures known as Graph eXchange Language (GXL)[4] format. GXL is designed to provide a standard exchange format for graphs [94]. It is an XML-based standard exchange format to share data between tools and it can also be extended to represent hypergraphs and hierarchical graphs. Other XML based graph file formats includes GraphML, DGML and XGMML. GraphML[5] is an XML-based comprehensive file format for graphs. It can represent different types of graphs, including directed, undirected, hypergraphs, and hierarchical graphs. Directed Graph Markup Language (DGML) is used for directed graphs. XGMML (eXtensible Graph Markup and Modeling Language) is an XML application based on GML, which is used for graph representation. Graph Modelling Language (GML) is a hierarchical text-based file format for describing graphs. Other text-based file formats for graphs are DOT and TGF. DOT is a graph description language. The DOT language only defines a graph; it does not provide support for rendering the graph. Trivial Graph Format (TGF) is a simple text-based file format for describing graphs using adjacency list data structure.

Graphviz[6] is an open-source graph visualization tool. It provides different graphical interfaces and support for the conversion of various graph file formats. Other graph visualization and analysis tools includes iGraph[7], Jgraph[8].

[1] `https://iapr-tc15.greyc.fr`

[2] `www.iapr.org`

[3] `https://iapr-tc15.greyc.fr/links.html`

[4] `http://www.gupro.de/GXL/`

[5] `http://graphml.graphdrawing.org/`

[6] `https://graphviz.org/`

[7] `https://igraph.org/`

[8] `https://sourceforge.net/projects/jgraph/`

DOI: 10.1201/9781003532248-11

There are several publicly available graph datasets and benchmarks[9] for evaluation and implementation of graph matching algorithms. IAM[10] Graph Database Repository is one of the publicly available graph databases for graph-based pattern recognition and machine learning. IAM graph database uses GXL file formats for graph representation and manipulation [148]. The ARG Database[11] is a large collection of labeled and unlabeled graphs provided by the MIVIA Group for benchmarking of graph matching algorithms. Graph Data Repository For Graph Edit Distance (GDR4GED)[12] provide datasets for testing graph edit distance algorithms. Benchmark Datasets[13] for Graph Kernels are provided evaluation and testing of graph kernels.

A novel Graph Matching Toolkit (GMT)[14] is a publicly available software toolkit, and it introduces a flexible software package for various implementation of graph edit distance computation [152]. GMT toolkit implements different algorithms of graph edit distance computation in JAVA. The VFLib[15] is a graph matching library which provides various algorithms for graph isomorphism, subgraph isomorphism [45]. The library is an efficient implementation of the VF2 graph matching algorithm and is written in C++. LAD[16] is a software toolkit for computing the subgraph isomorphism problem, and it is implemented in C. GEDLIB[17] is a library for finding edit distances between graphs using different algorithms, and it is written in C++. Another Graph Matching Toolkit (GMT)[18], provides C++ implementation of various graph matching algorithms. Graph-based tools for data mining and machine learning is provided in [23]. Overview of existing software tools for graph matching is described in [118].

[9] https://iapr-tc15.greyc.fr/links.html
[10] https://fki.tic.heia-fr.ch/databases/iam-graph-database
[11] https://mivia.unisa.it/datasets/graph-database/arg-database/
[12] http://www.rfai.li.univ-tours.fr/PublicData/GDR4GED/home.html
[13] https://ls11-www.cs.tu-dortmund.de/staff/morris/graphkerneldatasets
[14] http://www.fhnw.ch/wirtschaft/iwi/gmt
[15] https://mivia.unisa.it/vflib/
[16] http://liris.cnrs.fr/csolnon/LAD.html
[17] https://github.com/dbblumenthal/gedlib
[18] https://www3.cs.stonybrook.edu/~algorith/implement/gmt/implement.shtml

Bibliography

[1] C. C. Aggarwal and H. Wang. *Managing and mining graph data.* Advances in Database Systems, Springer, 2010.

[2] A. V. Aho, J. E. Hopcroft, and J. D. Ullman. *The design and analysis of computer algorithms.* Addison Wesley, 1974.w

[3] E. Akbas and P. Zhao. Graph clustering based on attribute-aware graph embedding. *From Security to Community Detection in Social Networking Platforms, Springer International Publishing, Cham*, pages 109–131, 2019.

[4] T. Akutsu, D. Fukagawa, A. Takasu, and T. Tamura. Exact algorithms for computing the tree edit distance between unordered trees. *Theoretical Computer Science*, 412:352–364, 2011.

[5] O. Allani, H. B. Zghal, N. Mellouli, and H. Akdag. A knowledge-based image retrieval system integrating semantic and visual features. *Procedia Computer Science*, 96:1428–1436, 2016.

[6] A. Armiti and M. Gertz. Vertex similarity – a basic framework for matching geometric graphs. *LWA*, pages 111–122, 2014.

[7] L. Babai and E. M. Luks. Canonical labeling of graphs. *Proceedings of the Fifteenth Annual ACM Symposium on Theory of Computing*, pages 171–183, 1983.

[8] L. Bai and E. R. Hancock. Graph kernels from the Jensen-Shannon divergence. *Journal of Mathematical Imaging and Vision*, 47:60–69, 2013.

[9] A. Barsi. Neural self-organization using graphs. *Int. Conf. on Machine Learning and data mining in Pattern Recognition, Lecture Notes in Artificial Intelligence, Springer*, 2734:343–352, 2003.

[10] S. Belongie and J. Malik. Matching with shape contexts. *Proceedings of IEEE Workshop on Content-Based Access of Image and Video Libraries*, pages 20–26, 2000.

[11] A. Bera, Z. Wharton, Y. Liu, N. Bessis, and A. Behera. SR-GNN: Spatial relation-aware graph neural network for fine-grained image categorization. *IEEE Transactions on Image Processing*, 31:6017–6031, 2022.

[12] M. Bernard, L. Boyer, A. Habrard, and M. Sebban. Learning probabilistic models for tree edit distance. *Pattern Recognition*, 41:2611–2629, 2008.

[13] S. Berretti, A. Del Bimbo, and E. Vicario. The computational aspect of retrieval by spatial arrangement. *Int. Conf. Pattern Recognition*, pages 1047–1051, 2000.

[14] S. Berretti, A. Del Bimbo, and E. Vicario. Efficient matching and indexing of graph models in content-based retrieval. *IEEE Transactions on Pattern Analysis and Machine Intelligence*, 23:1089–1105, 2001.

[15] P. Bille. A survey on tree edit distance and related problems. *Theoretical Computer Science*, 337:217–239, 2005.

[16] X. Bin, B. Jiajun, C. Chen, C. Wang, D. Cai, and X. He. EMR: a scalable graph-based ranking model for content-based image retrieval. *IEEE Transactions on Knowledge and Data Engineering*, 27(1):102–114, 2013.

[17] V. D. Blondel, J. Guillaume, R. Lambiotte, and E. Lefebvre. Fast unfolding of communities in large networks. *Journal of Statistical Mechanics: Theory and Experiment*, 2008(10):P10008, 2008.

[18] K. Borgwardt and H. Kriegel. Shortest-path kernels on graphs. *Int.Conf. Data Mining*, pages 74–81, 2005.

[19] F. Boubekeur, M. Boughanem, and L. Tamine-Lechani. Semantic information retrieval based on CP-nets. *IEEE International Fuzzy System Conference*, 121:1–7, 2007.

[20] M. Bouillon, R. Ingold, and M. Liwicki. Grayification: A meaningful grayscale conversion to improve handwritten historical document analysis. *Pattern Recognition Letters*, 121:46–51, 2019.

[21] L. Brun, P. Foggia, and M. Vento. Trends in graph-based representations for pattern recognition. *Pattern Recognition Letters*, https://doi.org/10.1016/j.patrec.2018.03.016, 2018.

[22] H. Bunke. Error-tolerant graph matching: a formal framework and algorithms. *Advances in Pattern Recognition, Int. Workshop on Structural, Syntactic and Statistical Pattern Recognition (SSPR and SPR), Lecture Notes in Computer Science, Springer*, 1451, 1998.

[23] H. Bunke. Graph-based tools for data mining and machine learning. *Proceedings of the 3rd Int. Conference on Machine learning and data mining in pattern recognition, LNAI*, 2734:7–19, 2003.

[24] H. Bunke and G. Allerman. Inexact graph matching for structural pattern recognition. *Pattern Recognition Letters*, 1:245–253, 1983.

[25] H. Bunke, P. Foggia, M. Vento, C. Sansone, and C. Guidobaldi. A comparison of algorithms for maximum common subgraph on randomly connected graphs. *Int. Workshop on Structural, Syntactic and Statistical Pattern Recognition (SSPR and SPR), Lecture Notes in Computer Science, Springer*, 2396, 2002.

[26] H. Bunke and K. Riesen. Towards the unification of structural and statistical pattern recognition. *Pattern Recognition Letters*, 33:811–825, 2012.

[27] R. Burkard, M. Dell'Amico, and S. Martello. *Assignment Problems*. Society for Industrial and Applied Mathematics, 2009.

[28] T. Caelli and S. Kosinov. Inexact graph matching using eigen-subspace projection clustering. *Int. Journal of Pattern Recognition and Artificial Intelligence*, 18:329–355, 2004.

[29] M. Carcaassoni and E. R. Hancock. Weighted graph-matching using modal clusters. *Int. Workshop on Structural, Syntactic and Statistical Pattern Recognition (SSPR and SPR), Lecture Notes in Computer Science, Springer*, pages 260–269, 2001.

[30] M. Chang and B. Kimia. Measuring 3D shape similarity by graph-based matching of the medial scaffolds. *Computer Vision and Image Understanding*, 115(5):707–720, 2011.

[31] H. Chen, A. Trouve, K. J. Murakami, and A. Fukuda. An intelligent annotation-based image retrieval system based on rdf descriptions. *Computers & Electrical Engineering*, 58:537–550, 2017.

[32] O. Cheong, J. Gudmundsson, H. Kim, D. Schymura, and F. Stehn. Measuring the similarity of geometric graphs. *Experimental Algorithms, Lecture Notes in Computer Science*, 5526:101–112, 2009.

[33] F. Chevalier, J. Domenger, J. Benois-Pineau, and M. Delest. Retrieval of objects in video by similarity based on graph matching. *Pattern Recognition Letters*, 28(8):939–949, 2007.

[34] W. Christmas, J. Kittler, and M. Petrou. Structural matching in computer vision using probabilistic relaxation. *IEEE Transactions on Pattern Analysis and Machine Intelligence*, 17:749–764, 1995.

[35] P. Chunanev. Community detection in node-attributed social networks: a survey. *Computer Science Review*, 37:1–24, 2020.

[36] N. D. Cilia, C. Stefano, F. Fontanella, and A. S. Freca. A ranking-based feature selection approach for handwritten character recognition. *Pattern Recognition Letters*, 121:77–86, 2019.

[37] D. Conte, P. Foggia, C. Sansone, and M. Vento. Thirty years of graph matching in pattern recognition. *Int. Journal of Pattern Recognition and Artificial Intelligence*, 18:265–298, 2004.

[38] D. J. Cook, N. Manocha, and L. B. Holder. Using a graph-based data mining system to perform web search. *Int. Journal of Pattern Recognition and Artificial Intelligence*, 17(5):705–720, 2003.

[39] S. A. Cook. The complexity of theorem-proving procedures. *Proceedings of Third ACM Symposium on Theory of Computing*, page 151–158, 1971.

[40] L. P. Cordella, P. Foggia, C. Sansone, F. Tortorella, and M. Vento. Graph matching:a fast algorithm and its evaluation. *Int. Conf. Pattern Recognition*, pages 1582–1584, 1998.

[41] L. P. Cordella, P. Foggia, C. Sansone, and M. Vento. An efficient algorithm for the inexact matching of ARG graphs using a contextual transformational model. *Int. Conf. Pattern Recognition*, pages 180–184, 1996.

[42] L. P. Cordella, P. Foggia, C. Sansone, and M. Vento. Subgraph transformations for the inexact matching of attributed relational graph. *GbRPR, Computing Supplement* 12:43–52, 1998.

[43] L. P. Cordella, P. Foggia, C. Sansone, and M. Vento. Performance evaluation of the VF graph matching algorithm. *Int. Conf. Image Analysis and Processing*, pages 1172–1177, 1999.

[44] L. P. Cordella, P. Foggia, C. Sansone, and M. Vento. An improved algorithm for matching large graphs. *Int. Workshop on Structural, Syntactic and Statistical Pattern Recognition (SSPR and SPR), Lecture Notes in Computer Science, Springer*, page 149–159, 2001.

[45] L. P. Cordella, P. Foggia, C. Sansone, and M. Vento. A (sub)graph isomorphism algorithm for matching large graphs. *IEEE Transactions on Pattern Analysis and Machine Intelligence*, 26(10):1367–1372, 2004.

[46] T. Cour, P. Srinivasan, and J. Shi. Balanced graph matching. *Advances in Neural Information Processing System*, 19:313–322, 2007.

[47] T. Cover. Geometrical and statistical properties of system of linear inequality with applicatios in pattern recognition. *IEEE Transactions on Electronic Computers*, 14:326–334, 1965.

[48] N. Dahm, H. Bunke, T. Caelli, and Gao Y. Topological features and iterative node elimination for speeding up subgraph isomorphism detection. *Int. Conf. Pattern Recognition*, pages 1164–1167, 2012.

[49] F. J. Damerau. A technique for computer detection and correction of spelling errors. *Communications of the ACM*, 7(3):171–176, 1964.

[50] P. T. Darga, M. H. Liffiton, K. A. Sakallah, and Markov I. L. Exploiting structure in symmetry detection for CNF. *Design and Automation Conference*, pages 530–534, 2004.

[51] P. T. Darga, K. A. Sakallah, and Markov I. L. Faster symmetry discovery using sparsity of symmetries. *Design and Automation Conference*, pages 149–154, 2008.

[52] M. De Santo, P. Foggia, C. Sansone, and Vento M. A large database of graphs and its use for benchmarking graph isomorphism algorithms. *Pattern Recognition Letters*, 24:1067–1079, 2003.

[53] P. J. Dickinson, H. Bunke, A. Dadej, and M. Kraetzl. Matching graphs with unique node labels. *Pattern Analysis and Applications*, 7:243–254, 2004.

[54] O. Duchenne, F. Bach, I. Kweon, and J. Ponce. A tensor-based algorithm for high order graph matching. *IEEE Transactions on Pattern Analysis and Machine Intelligence*, 33(12):2383–2395, 2011.

[55] R. Duda, P. Hart, and D. Stork. *Pattern Classification.* Wiley-Interscience, 2000.

[56] S. P. Dwivedi. *Some algorithms on exact, approximate and error-tolerant graph matching.* Ph.D. Thesis, Indian Institute of Technology (BHU) Varanasi, Bharat, (arXiv:2012.15279), 2020.

[57] S. P. Dwivedi. Inexact graph matching using centrality measures. *Computing Research Repository, arXiv:2201.04563*, pages 1–12, 2021.

[58] S. P. Dwivedi. Approximate bipartite graph matching by modifying cost matrix. *Lecture Notes in Electrical Engineering,* Springer, 837:415–422, 2022.

[59] S. P. Dwivedi. Approximate grap edit distance for structural pattern recognition. *NeuroQuantology*, 20(14):798–804, 2022.

[60] S. P. Dwivedi and R. S. Singh. Error-tolerant graph matching using homeomorphism. *Int. Conf. on Advances in Computing, Communication and Informatics (ICACCI)*, pages 1762–1766, 2017.

[61] S. P. Dwivedi and R. S. Singh. Error-tolerant geometric graph similarity. *Int. Workshop on Structural, Syntactic and Statistical Pattern Recognition (SSPR and SPR), Lecture Notes in Computer Science, Springer*, 11004:337–344, 2018.

[62] S. P. Dwivedi and R. S. Singh. Error-tolerant graph matching using node contraction. *Pattern Recognition Letters*, 116:58–64, 2018.

[63] S. P. Dwivedi and R. S. Singh. Error-tolerant geometric graph similarity and matching. *Pattern Recognition Letters*, 125:625–631, 2019.

[64] S. P. Dwivedi and R. S. Singh. Error-tolerant approximate graph matching utilizing node centrality information. *Pattern Recognition Letters*, 133:313–319, 2020.

[65] S. P. Dwivedi and R. S. Singh. Pattern recognition using graph edit distance. *Sustainable Computing,* Springer, pages 335–349, 2023.

[66] S. P. Dwivedi, V. Srivastava, and U. Gupta. Graph similarity using tree edit distance. *Int. Workshop on Structural, Syntactic and Statistical Pattern Recognition (SSPR and SPR), Lecture Notes in Computer Science,* Springer, 13813:233–241, 2022.

[67] M. A. Eshera and K. S. Fu. A graph distance measure for image analysis. *IEEE Transactions on Systems, Man, and Cybernetics*, 14(3):398–408, 1984.

[68] P. F. Felzenszwalb and D. P. Huttenlocher. Efficient graph-based image segmentation. *Intl. Journal of Computer Vision*, 59(2):167–181, 2004.

[69] M. Ferrer, F. Serratosa, and K. Riesen. Improving bipartite graph matching by assessing the assignment confidence. *Pattern Recognition Letters*, 65:29–36, 2015.

[70] M. A. Fischler and R. A. Elschlager. The representation and matching of pictorial structures. *IEEE Transactions on Computers*, 22:67–92, 1973.

[71] P. Foggia, G. Percannella, and M. Vento. Graph matching and learning in pattern recognition in the last 10 years. *Int. Journal of Pattern Recognition and Artificial Intelligence*, 28:1450001.1–1450001.40, 2014.

[72] P. Frasconi, M. Gori, and A. Sperduti. A general framework for adaptive processing of data structures. *IEEE Transactions on Neural Networks*, 9:768–786, 1998.

[73] L. C. Freeman. A set of measures of centrality based upon betweenness. *Sociometry*, 40:35–41, 1977.

[74] L. C. Freeman. *The Development of Social Network Analysis.* Empirical Press, Vancouver, 2004.

[75] K. S. Fu and B. K. Bhargava. Tree systems for syntactic pattern recognition. *IEEE Transactions on Computers*, 22:1087–1099, 1973.

[76] X. Gao, B. Xiao, D. Tao, and X. Li. A survey of graph edit distance. *Pattern Analysis and Applications*, 13:113–129, 2010.

[77] M. R. Garey and D. S. Johnson. *Computers and Intractability: A Guide to the Theory of NP-Completeness.* W. H. Freeman and Company, 1979.

[78] T. Gartner. *Kernels for structured data.* World Scientific, 2008.

[79] B. Gazere, L. Brun, and D. Villemin. Two new graphs kernels in chemoinformatics. *Pattern Recognition Letters*, 33(15):2038–2047, 2012.

[80] D. E. Ghahraman, A. K. Wong, and T. Au. Graph optimal monomorphism algorithms. *IEEE Transactions on Systems, Man, and Cybernetics*, 10(4):181–188, 1980.

[81] M. Girvan and M. E. J. Newman. Community structure in social and biological network. *Applied Mathematics*, 99(12):7821–7826, 2002.

[82] S. Gold and A. Rangarajan. A graduated assignment algorithm for graph matching. *IEEE Transactions on Pattern Analysis and Machine Intelligence*, 18(4):377–388, 1996.

[83] F.M.F. Gonçalves, I.R. Guilherme, and D.C.G. Pedronette. Semantic guided interactive image retrieval for plant identification. *Expert System Applications*, 91:12–26, 2018.

[84] M. Gori, M. Maggini, and L. Sarti. Exact and approximate graph matching using random walks. *IEEE Transactions on Pattern Analysis and Machine Intelligence*, 27(7):1100–1111, 2005.

[85] L. Grady. Random walks for image segmentation. *IEEE Transactions on Pattern Analysis and Machine Intelligence*, 28(11):01–17, 2006.

[86] L. Gregory and J. Kittler. Using graph search techniques for contextual colour retrieval. *Int. Workshop on Structural, Syntactic and Statistical Pattern Recognition (SSPR and SPR), Lecture Notes in Computer Science*, Springer, pages 186–194, 2002.

[87] K. Han, Y. Wang, J. Guo, Y. Tang, and E. Wu. Vision gnn: an image is worth graph of nodes. *Int. Conf. Neural Information Processing Systems*, pages 1–13, 2022.

[88] E. R. Hancock and J. Kittler. Discrete relaxation. *Pattern Recognition*, 23:711–733, 1990.

[89] E. R. Hancock and R. C. Wilson. Pattern analysis with graphs: parallel work at bern and yark. *Pattern Recognition Letters*, 33:833–841, 2012.

[90] P. E. Hart, N. J. Nilson, and B. Raphael. A formal basis for heuristic determination of minimum cost paths. *IEEE Transactons on System, Science and Cybernetics*, 4:100–107, 1968.

[91] D. Haussler. *Convolution kernels on discrete structures.* University of California, Santa Cruz, Technical Report UCSC-CRL-99-10, 1999.

[92] D. Hidovic and M. Pelilo. Metrics for attributed graphs based on the maximal similarity common subgraph. *Int. Journal of Pattern Recognition and Artificial Intelligence*, 28(3):299–313, 2004.

[93] C. M. Hoffmann and M. O'Donnel. Pattern matching in trees. *Journal of the Association for Computing Machinery*, 29(1):68–95, 1982.

[94] R. C. Holt, A. Schürr, S. E. Sim, and A. Winter. GXL: A graph-based standard exchange format for reengineering. *Science of Computer Programming*, 60(2):149–170, 2006.

[95] R. C. Holt, A. Winter, and A. Schürr. GXL: Towards a standard exchange format. *Proceedings of 7th Working Conference on Reverse Engineering*, 2000.

[96] J. E. Hopcroft and J. Wong. Linear time algorithm for isomorphism of planar graphs. *ACM Symposium on Theory of Computing*, pages 172–184, 1974.

[97] B. Huet and E. R. Hancock. Shape recognition from large image libraries by inexact graph matching. *Pattern Recognition Letters*, 20:1259–1269, 1999.

[98] C. Irniger and H. Bunke. Decision tree structures for graph database filtering. *Int. Workshop on Structural, Syntactic and Statistical Pattern Recognition (SSPR and SPR), Lecture Notes in Computer Science, Springer*, page 66–75, 2004.

[99] A. Jagota, M. Pelillo, and A. Rangarajan. A new deterministic annealing algorithm for maximum clique. *Int. Joint Conf. Neural Networks*, pages 505–508, 2000.

[100] A. K. Jain, K. Nandakumar, and A. Ross. 50 years of biometric research: Accomplishments, challenges, and opportunities. *Pattern Recognition Letters*, 79:80–105, 2016.

[101] B. Jain and K. Obermayer. Graph quantization. *Computer Vision and Image Understanding*, 115:946–961, 2011.

[102] B. Jain and F. Wysotzki. Solving inexact graph isomorphism problems using neural networks. *Neurocomputing*, 63:45–67, 2005.

[103] K. Jenni, S. Mandala, and Sunar M. S. Content based image retrieval using colour strings comparison. *Procedia Computer Science*, 50:374–379, 2015.

[104] J. Johnson, R. Krishna, Stark M., L. Li-jia, D. Shamma, M. Bernstein, and L. Fie. Image retrieval using scene graphs. *Proceedings of the IEEE Conference on Computer Vision and Pattern Recognition*, pages 3668–3678, 2015.

[105] K. Kashima, K. Tsuda, and A. Inokuchi. Marginalized kernels between labeled graphs. *Int. Conf. Machine Learning*, 20:321–328, 2003.

[106] B. Khemani, S. Patil, K. Kotecha, and S. Tanwar. A review of grap neural networks: concepts, architectures, techniques, challenges, datasets, applications, and future directions. *Jouranl of Big Data*, 11:1–43, 2024.

[107] K. G. Khoo and P. N. Suganthan. Multiple relational graphs mapping using genetic algorithms. *Proc. Congress on Evolutionary Computation*, pages 727–733, 2001.

[108] J. Kittler and E. R. Hancock. Combining evidence in probabilistic relaxation. *Int. Journal of Pattern Recognition and Artificial Intelligence*, 3:29–51, 1989.

[109] U. Kobler, J. Schoning and J. Torjan. *The Graph Isomorphism Problem: Its Structural Complexity*. Progress in Theoretical Computer Science, Birkhäuser Basel, 1993.

[110] I. Koch. Enumerating all connected maximal common subgraphs in two graphs. *Theoretical Computer Science*, 250:1–30, 2001.

[111] J. Konc and D. Janezic. An improved branch and bound algorithm for maximum clique problem. *MATCH Communications in Mathematical and in Computer Chemistry*, 58:569–590, 2007.

[112] J. H. Koo and S. I. Yoo. A structural matching for two-dimensional visual pattern insepction. *Proceedings of the IEEE Intl. Conf. on Systems, Man and Cybernetics*, pages 4429–4434, 1998.

[113] A. Kostin, J Kittler, and W. Christmas. Object recognition by symmetrised graph matching using relaxation labelling with an inhibitory mechanism. *Pattern Recognition Letters*, 26:381–393, 2005.

[114] M. Kuramochi and Karypis G. Discovering frequent geometric subgraphs. *Information Systems*, 32:1101–1120, 2007.

[115] R. E. Ladner. On the structure of polynomial time reducibility. *Journal of the Association for Computing Machinery*, 22(1):155–171, 1975.

[116] J. Lafferty and G. Lebanon. Diffusion kernels on statistical manifolds. *Journal of Machine Learning Research*, 6:129–163, 2005.

[117] J. Larrosa and G. Valiente. Constraint satisfaction algorithms for graph pattern matching. *Mathematical Structures in Computer Science*, 12:403–422, 2002.

[118] K. Lasinger. Overview of existing software tools for graph matching. *Technical Report, PRIP-TR-130*, 2013.

[119] V. I. Levenshtein. Binary codes capable of correcting deletions, insertions and reversals. *Soviet Physics Doklady*, 10:707–710, 1966.

[120] C. Y. Li and C. T. Hsu. Image retrieval with relevance feedback based on graph-theoretic region correspondence estimation. *IEEE Transactions on Multimedia*, 10(3):447–456, 2008.

[121] W. J. Li and T. Lee. Object recognition by sub-scene graph matching. *IEEE Int. Conf. on Robotics and Automation*, pages 1459–1464, 2000.

[122] C. W. Liu, K. C. Fan, J. T. Horng, and Y. K. Wang. Solving weighted graph matching problem by modified microgenetic algorithm. *Int. Conf. on Systems Man Cybernetics*, pages 638–643, 1995.

[123] L. Livi and A. Rizzi. The graph matching problem. *Pattern Analysis and Applications*, 16:253–283, 2013.

[124] J. Llados, E. Marti, and J. Villnueva. Symbol recognition by error-tolerant subgraph matching between region adjacency graphs. *IEEE Transactions on Pattern Analysis and Machine Intelligence*, 23:1137–1143, 2001.

[125] S. Y. Lu. A tree matching algorithm based on node splitting and merging. *IEEE Transactions on Pattern Analysis and Machine Intelligence*, 6(2):249–256, 1984.

[126] E. M. Luks. Isomorphism of graphs of bounded valence can be tested in polynomial time. *Journal of Computer and System Sciences*, 25:42–65, 1982.

[127] B. Luo and E. R. Hancock. Structural graph matching using the EM algorithm and singular value decomposition. *IEEE Transactions on Pattern Analysis and Machine Intelligence*, 23(10):1120–1136, 2001.

[128] L. Maisonnasse, J. P. Chevallet, C. Berrut. Incomplete and fuzzy conceptual graph to automatically index medical reports. *Lecture Notes in Computer Science, Springer*, 4592:240–251, 2007.

[129] J. J. McGregor. Backtrack search algorithm and the maximal common subgraph problem. *Software — Practice and Experience*, 12:23–34, 1982.

[130] B. D. McKay. Practical graph isomorphism. *Congressus Numerantium*, 30:45–87, 1981.

[131] B. D. McKay and A. Piperno. Practical graph isomorphism II. *Journal of Symbolic Computation*, 60:94–112, 2014.

[132] M. Mechkour, C. Berrut, and Y. Chiaramella. Using conceptual graph frame work for image retrieval. *International Conference on MultiMedia Modeling*, pages 127–142, 1995.

[133] B. Messmer and H. Bunke. A decision tree approach to graph and subgraph isomorphism detection. *Pattern Recognition*, 32:1979–1998, 1998.

[134] B. Messmer and H. Bunke. Efficient subgraph isomorphism detection: a decomposition approach. *IEEE Transactions on Knowledge and Data Engineering*, 12:307–323, 2000.

[135] M. Mori, S. Uchida, and H. Sakano. Global feature for online character recognition. *Pattern Recognition Letters*, 35:142–148, 2014.

[136] R. Myers, R. C. Wilson, and E. R. Hancock. Bayesian graph edit distance. *IEEE Transactions on Pattern Analysis and Machine Intelligence*, 22:628–635, 2000.

[137] M. Neuhaus and H. Bunke. An error-tolerant approximate matching algorithm for attributed planar graphs and its application to fingerprint classification. *Int. Workshop on Structural, Syntactic and Statistical Pattern Recognition (SSPR and SPR), Lecture Notes in Computer Science,* Springer, 3138:180–189, 2004.

[138] M. Neuhaus and H. Bunke. *Bridging the gap between graph edit distance and kernel machines.* World Scientific, 2007.

[139] M. Neuhaus, K. Riesen, and H. Bunke. Fast suboptimal algorithms for the computation of graph edit distance. *Int. Workshop on Structural, Syntactic and Statistical Pattern Recognition (SSPR and SPR), Lecture Notes in Computer Science,* Springer, 4109:163–172, 2006.

[140] M. Neuhaus, K. Riesen, and H. Bunke. Novel kernels for error-tolerant graph classification. *Spatial Vision*, 22(5):425–441, 2009.

[141] M.E.J. Newman. *Networks–An Introduction.* Oxford University Press, 2010.

[142] N. J. Nilsson. *Principles of artificial intelligence.* Springer, 1980.

[143] K. Oflazer. Error-tolerant retrieval of trees. *IEEE Transactions on Pattern Analysis and Machine Intelligence*, 19:1376–1380, 1997.

[144] E. Pekalska and R. Duin. *The dissimilarity representation for pattern recognition: Foundations and applications.* World Scientific, 2005.

[145] M. A. Pinheiro, J. Kybic, and P. Fua. Geometric graph matching using monte carlo tree search. *IEEE Transactions on Pattern Analysis and Machine Intelligence*, 39(11):2171–2185, 2017.

[146] A. Rangarajan and E. D. Mjolsness. A lagrangian relaxation network for graph matching. *IEEE Transactions on Neural Networks*, 7:1365–1381, 1996.

[147] K. Riesen. *Structural Pattern Recognition with Graph Edit Distance, Approximation Algorithms and Applications.* Springer, 2015.

[148] K. Riesen and H. Bunke. IAM graph database repository for graph based pattern recognition and machine learning. *Int. Workshop on Structural, Syntactic and Statistical Pattern Recognition (SSPR and SPR), Lecture Notes in Computer Science,* Springer, 5342:287–297, 2008.

[149] K. Riesen and H. Bunke. Approximate graph edit distance computation by means of bipartite graph matching. *Image and Vision Computing,* 27:950–959, 2009.

[150] K. Riesen and H. Bunke. *Graph Classification and Clustering based on Vector Space Embedding.* World Scientific, 2010.

[151] K. Riesen and H. Bunke. Improving bipartite graph edit distance approximation using various search strategies. *Pattern Recognition,* 48:1349–1363, 2015.

[152] K. Riesen, S. Emmenegger, and H. Bunke. A novel software toolkit for graph edit distance computation. *Int. Workshop on Graph Based Representations in Pattern Recognition, Lecture Notes in Computer Science, Springer,* 7877:142–151, 2013.

[153] K. Riesen, A. Fischer, and H. Bunke. Estimating graph edit distance using lower and upper bounds of bipartite approximations. *Int. Journal of Pattern Recognition and Artificial Intelligence,* 29:1550011.1–1550011.27, 2015.

[154] A. Robles-Kelly and E. R. Hancock. Graph edit distance from spectral seriation. *IEEE Transactions on Pattern Analysis and Machine Intelligence,* 27:365–378, 2005.

[155] J. Rocha and T. Pavlidis. A shape analysis model with application to character recognition sysytem. *IEEE Transactions on Pattern Analysis and Machine Intelligence,* 16:393–404, 1994.

[156] K. Rosen. *Discrete Mathematics and Its Applications.* McGraw Hill Education, 2017.

[157] A. Rosenfeld, R. A. Hummel, and S. W. Zucker. Scene labeling by relaxation operations. *IEEE Transactons on System, Man and Cybernetics,* 6:420–433, 1976.

[158] Y. Rui, T. S. Huang, M. Ortega, and Mehrotra S. Relevance feedback: a power tool for interactive content-based image retrieval. *IEEE Transactions on Circuit and Systems for Video Technology,* 8(5):644–655, 1998.

[159] A. Sanfeliu, R. Alquezar, J. Andrade, J. Climent, F. Serratosa, and J. Verges. Graph-based representations and techniques for image processing and image analysis. *Pattern Recognition*, 35(3):639–650, 2002.

[160] A. Sanfeliu and K. S. Fu. A distance measure between attributed relational graphs for pattern recognition. *IEEE Transactons on System, Man and Cybernetics*, 13:353–363, 1983.

[161] A. Sanfeliu, F. Serratosa, and R. Alquezar. Second-order random graphs for modeling sets of attributed graphs and their application to object learning and recognition. *Int. Journal of Pattern Recognition and Artificial Intelligence*, 18(3):375–396, 2004.

[162] G. Sanroma, R. Alquezar, and F. Serratosa. A new graph matching method for point-set correspondence using the EM algorithm and softassign. *Computer Vision and Image Understanding*, 116(2):292–304, 2012.

[163] P. Santacruz and F. Serratosa. Learning the sub-optimal graph edit distance edit costs based on an embedded model. *Int. Workshop on Structural, Syntactic and Statistical Pattern Recognition (SSPR and SPR), Lecture Notes in Computer Science, Springer*, page 282–292, 2018.

[164] K. C. Santosh. *Document Image Analysis: Current Trends and Challenges in Graphics Recognition.* Springer Singapore, 2018.

[165] F. Scarselli, M. Gori, A. C. Tsoi, M. Hagenbuchner, and G. Monfardini. The graph neural network model. *IEEE Transactions on Neural Network*, 20:71–77, 2009.

[166] B. Scholkopf and A. Smola. *Learning with kernels.* MIT Press, 2002.

[167] T. B. Sebastian, P. N. Klein, and B. B. Kimia. Recognition of shapes by editing their shock graphs. *IEEE Transactions on Pattern Analysis and Machine Intelligence*, 26(5):550–571, 2005.

[168] F. Serratosa, R. Alquezar, and A. Sanfeliu. Synthesis of function-described graphs and clustering of attributed graphs. *Int. Journal of Pattern Recognition and Artificial Intelligence*, 16(6):621–655, 2002.

[169] F. Serratosa, R. Alquezar, and A. Sanfeliu. Function-described graphs for modelling objects represented by sets of attributed graphs. *Pattern Recognition*, 36(3):781–798, 2003.

[170] F. Serratosa, R. Alquezer, and A. Sanfeliu. Function-described graphs: a fast algorithm to compute a sub-optimal matching measure. *Int. Workshop on Graph-Based Representations in Pattern Recognition*, pages 71–77, 1999.

[171] L. G. Shapiro and R. M. Haralick. Structural descriptions and inexact matching. *IEEE Transactons on Pattern Analysis and Machine Intelligence*, 3:504–519, 1981.

[172] D. Shasha, J. L. Wang, k. Zhang, and F.Y. Shih. Exact and approximate algorithms for unordered tree matching. *IEEE Transactons on Systems, Man and cybernetics*, 24:668–678, 1994.

[173] J. Shawe-Taylor and N. Cristianini. *Kernel Methods for Pattern Analysis.* Cambridge University Press, 2004.

[174] J. Shi and J. Malik. Normalized cuts and image segmentation. *IEEE Transactions on Pattern Analysis and Machine Intelligence*, 22(8): 888–905, 2000.

[175] J. Shieh, Y. Yeh, C. Lin, and J. Wu. Collaborative knowledge semantic graph image search. *Proceedings of the 17th International Conference on World Wide Web*, pages 1055–1056, 2008.

[176] A. Shokoufandeh, D. Macrini, S. Dickinson, K. Siddiqi, and S. Zucker. Indexing hierarchical structures using graph spectra. *IEEE Transactions on Pattern Analysis and Machine Intelligence*, 27:1125–1140, 2005.

[177] A. Sole-Ribalta, F. Serratosa, and A. Sanfeliu. On the grap edit distance cost: properties and applications. *Int. Journal of Pattern Recognition and Artificial Intelligence*, 26(5):1260004.1–1260004.21, 2012.

[178] C. Solnon. All different-based filtering for subgraph isomorphism. *Artificial Intelligence*, 174:850–864, 2010.

[179] C. Solnon. Experimental evaluation of subgraph isomorphism solvers. *Int. Workshop on Graph Based Representations in Pattern Recognition, Lecture Notes in Computer Science,* Springer, 11510:1–13, 2019.

[180] A. Sperduti and A. Starita. Supervised neural networks for the classification of structures. *IEEE Transactions on Neural Networks*, 8:714–735, 1997.

[181] B. Strug. Using kernels on hierarchical graphs in automatic classification of designs. *Int. Workshop on Graph Based Representations in Pattern Recognition, Lecture Notes in Computer Science,* Springer, 6658: 335–344, 2011.

[182] P. Suganthan, E. Teoh, and D. Mital. Pattern recognition by homomorphic graph matching using hopfield neural network. *Image and Vision Computing*, 13:45–60, 1995.

[183] A. Torsello and E. R. Hancock. Computing approximate tree edit distance using relaxation labeling. *Pattern Recognition Letters*, 24:1089–1097, 2003.

[184] W. H. Tsai and K. S. Fu. Error-correcting isomorphisms of attributed relational graphs for pattern analysis. *IEEE Transactons on System, Man and Cybernetics*, 9:757–768, 1979.

[185] W. H. Tsai and K. S. Fu. Subgraph error-correcting isomorphism for syntactic pattern recognition. *IEEE Transactons on System, Man and Cybernetics*, 13:48–62, 1983.

[186] J. R. Ullman. An algorithm for subgraph isomorphism. *Journal of the Association for Computing Machinery*, 23(1):31–42, 1976.

[187] J. R. Ullman. Bit-vector algorithms for binary constraint satisfaction and subgraph isomorphism. *ACM Journal of Experimental Algorithmics*, 15(1):1.6.1–1.6.64, 2011.

[188] S. Umeyama. An eigendecomposition approach to weighted graph matching problem. *IEEE Transactions on Pattern Analysis and Machine Intelligence*, 10(5):695–703, 1988.

[189] G. Valiente. An efficient bottom up distance between trees. *Int. Symp. String Processing and Information Retrieval*, pages 212–219, 2001.

[190] V. Vapnik. *Statistical Learning Theory.* John Wiley, 1998.

[191] C. Wang and K. Abe. Region correspondence by inexact attributed planar graph matching. *Int. Conf. Computer Vision*, pages 440–447, 1995.

[192] J. Wang, K. Zhang, and G. Chirn. The approximate graph matching problem. *Int. Conf. Pattern Recognition*, pages 284–288, 1994.

[193] M. Weber, M. Liwicki, and A. Dengel. Indexing with well-founded total order for faster subgraph isomorphism detection. *Int. Workshop on Graph Based Representations in Pattern Recognition, Lecture Notes in Computer Science,* Springer, 6658:185–194, 2011.

[194] R. Wilson and E. Hancock. Structural matching by discrete relaxation. *IEEE Transactions on Pattern Analysis and Machine Intelligence*, 19:634–648, 1997.

[195] Y. Wu, J. Lim, and M. H. Yang. Object tracking benchmark. *IEEE Transactions on Pattern Analysis and Machine Intelligence*, 37(9):1834–1848, 2015.

[196] S. Xiao, J. Li, J. Lu, S. Huang, B. Zeng, and S. Wang. Graph neural networks for multi-view learning: a taxonomic review. *Artificial Intelligence Review*, 57:1–39, 2024.

[197] H. Yang, Y. Li, W. Li, X. Wang, and F. Yang. Content-based image retrieval using local visual attention feature. *Journal of Visual Communication and Image Representation*, 25(6):1308–1323, 2014.

[198] S. Zampelli, Y. Deville, and C. Solnon. Solving subgraph isomorphism problems with constraint programming. *Constraints*, 15:327–353, 2010.

[199] V. N. Zemlyachenko, N. M. Korneenko, and R. I. Tyshkevich. Graph isomorphism problem. *Journal of Mathematical Sciences*, 29(4):1426–1481, 1985.

[200] Z. Zeng, A.K.H. Tung, J. Wang, J. Feng, and L. Zhou. Comparing stars: on approximating graph edit distance. *PVLDB*, 2:25–36, 2009.

[201] K. Zhang, R. Statman, and D. Shasha. On the editing distance between unordered labeled trees. *Information Processing Letters*, 42(3):133–139, 1992.

[202] X. Zhong, W. Kuang, and Z. Li. Adaptive graph reasoning network for object detection. *Image and Vision Computing*, 151:105–248, 2024.

Index

Made in the USA
Las Vegas, NV
19 September 2026

a88cade3-5e30-4371-b142-c818ff5f2098R02